# TALES
# OF THE
# CHESHIRE
# PLANES

*John McDaniel*

*TALES OF THE CHESHIRE PLANES*

First published 1998
by GMS Enterprises
67 Pyhill, Bretton, Peterborough,
England PE3 8QQ
Tel and Fax (01733) 265123
EMail: GMSAVIATIONBOOKS@ Compuserve.com


ISBN: 1 870384  64 4

Printed and bound for GMS Enterprises

# CONTENTS

# Acknowledgements

Firstly, I would like to feel that the writing of this book allows me to express my gratitude to all my colleagues in the Flight Development Department and the Test Pilots Section at Woodford airfield, from 1947 to 1978, who shared with me an enjoyable and, at times, a quite exciting career.

In addition to those in AVRO/HSA/BAe, I want to say a very special 'thank you' to my very good friend Gordon Fox, the Senior Rolls-Royce Representative at Woodford throughout this period. When it came down to the 'nuts and bolts' and the related problems of aero engines, there was no one better. Many of the experiences related in this book I shared with Gordon, and without his help with the text and photographs, I probably would not have attempted to write it.

Particularly when recalling episodes in the testing and development of the Vulcan, my special thanks to Jimmy Harrison for prompting my fading memory and for the loan of photographs. Jim was one of our great Chief Test Pilots, all of whom I thank for the privilege of flying and working with.

My thanks also to Peter V. Clegg for his advice that set me off on the right track and, in passing, for writing that wonderful book on the life of a man whom I respected very much, Jimmy Orrell, a past Chief Test Pilot.

Finally my wife Joan, who for some years was a member of the team at Woodford, for her contribution to the book (Lucy's Treat), her diligent proof reading and for her support and encouragement.

# The Author...

John McDaniel entered the RAF in 1940 via the Apprentice School at Halton. As an 'ex-brat' Fitter IIA, he helped keep Wellingtons in the air for a couple of years, including sending them off on the famous 1000 bomber raids. Eventually he received his posting for pilot training but it was not until 1945 that he joined what turned out to be the last course in Canada on the Empire Aircrew Training Scheme.

After the war, Roy Dobson took him on as a draughtsman in the A. V. Roe Design Office. After six months, John was moved into the newly formed Flight Test Section at Woodford at the request of Sqn.Ldr. D.Wilson who had just arrived from 617 Squadron to form the new section. Sadly, Sqn. Ldr. Wilson was killed in the Tudor II accident that year.

By 1960, the section of only two staff in 1947 had become the Flight Development Department with a staff of 120, the expansion being mainly to cope with the development and testing of the Vulcan bomber.

*The author (left), along with Test-Pilot Ken Cook (white overalls) and Flight Engineers Eric Allen and Don Wilson about to take up Tudor IV 'Star Leopard' (G-AHNN) for a test-flight at Woodford prior to delivery to British South American Airways in June 1948.*

## TALES OF THE CHESHIRE PLANES

In 1978, after 31 years of flight testing at Woodford, John accepted a move to BAe Kingston and went out to Egypt to join a team overhauling MiG21's.

Retiring in 1986, he and his close friend Gordon Fox, formed a partnership as engineering consultants using the name Dartech. This business ran successfully for about five years and provided further interesting work and travel in various parts of the World.

Until he also retired, Gordon Fox was the Senior Rolls-Royce Representative at Woodford and the success of flight development there depended very much on Gordon's skill and dedicated hard work, that ensured that the R-R engines were always at the correct standard and fully serviceable for the ground and flight tests to be carried out.

Gordon's opinions and his carefully considered judgements were listened to and valued by ground and flight staff alike and none more so than by the author.

# Introduction

Many stories of life's experiences are told over a pint at the bar, or holding a glass at the dinner table.

The author of this book is among the fortunate who can 'tell a tale or two'. After hearing listeners say, so many times, *"you should write a book..."*, what you are holding is the result.

Aviation development has had many exciting moments, starting with the Wright Brothers but the greatest period must surely be the post-war years of the 1950's, '60's, and '70's. This book relates to aircraft and engine development during this period, but rather than just swamping the reader with the technical story, it recalls lighthearted aspects of life along the way. The author is the first to admit how lucky he was to have been part of it and in these days of the silicon chip and the terrorist bomb, who would not want to relive those days again - the author certainly would.

The tales related are not told in any logical or historical sequence but rather as 'off the cuff' reminiscences as might come up naturally in a reunion or an evening with old friends. We hope that the book is of interest not only to the older generation but also, to say, younger persons embarking on their first overseas holiday on a charter flight. They might like some reading which has some relevance to the wonderful piece of aeronautical engineering that is carrying them upwards and onwards to sunnier climes.

However, I do not want any reader to imagine the life of test flying was all laughs, beer and skittles. There were tears too, hard work and sometimes sleepless nights, for one reason or another.

## CHAPTER ONE

# The Pee Tube

Returning from Canada on the old *Ile de France* in 1945, resplendent in uniform bearing sergeant's stripes and pilot's wings, was like a five day holiday cruise. After all the anxious waiting to get on the pilot's course; a year at Heaton Park in Manchester; flying training which stopped when the war in Europe ended and restarted after two days of sweating in the crew room; and the hard work and the worry as to whether or not I would be thrown off the course when I ground-looped on landing - this was now all behind me. As we ploughed across the Atlantic, which hopefully was now empty of German U-boats, I contemplated the good life that lay ahead. A King's commission and some real aeroplanes to fly awaited, or so I thought! The commission and a trip to Gieves materialised, but that was all. The real action in the RAF was on hold and the surviving front line men were taking a well earned stand down and rest.

I was posted to some station (its whereabouts today, I cannot even recall!) No duties to speak of and all week-ends were long ones at home, not an aircraft to be seen let alone to fly. After a few months of this frustrating inactivity I left the RAF and I don't recall the hierarchy shedding any tears, it being one less mouth to feed and one less monthly cheque into Lloyds Bank. I went to live in Lancashire and wrote to Roy Dobson at A.V. Roe asking for a job. He passed the letter down to the Asst. Chief Draughtsman, Eddie Dean, with a note on the bottom: *"Give this lad a chance!"*

A chance I was given and I entered the huge Drawing Office at Chadderton and was given a board on the air conditioning section concerned with the Tudor airliners. The Section Leader gave me a job and this resulted in me receiving one of most important lessons of my life from one the greatest aircraft designers this country has ever had - a lesson I have never forgotten to this day.

The job I was given was to design a 'Pee Tube'. Apparently, Tudors were being sold to a South American airline and top of the list of the special

features the airline's chief pilot required was a device beside his seat to allow him to relieve himself without having to get out of his seat - perhaps he had some prostate problem! No doubt I had been given the simplest job on the list. I could mend an aircraft, I could fly one but no one had told me how to design one. I imagined it was just a matter of drawing but soon I was to find out how little I knew.

*Avro Test Pilot Ken Cook airborne in the BOAC flagship Tudor I G-AGRF on 17th January 1947. The aircraft is en-route to London Airport where it was christened 'Elizabeth of England' by Princess Elizabeth on 21st January .*

I had just pinned a beautiful piece of blue linen drawing cloth on to my board and drawn a very neat chain dot centre line down the middle, when I heard a metallic tapping noise from further down the office and all went suddenly quiet. I subsequently learned that the first person to spot the Chief Designer, Roy Chadwick, entering the office would tap the metal lampshade over his desk as a warning signal for all to get their heads down and get working, or at least, give a reasonable impression of being busy. It did not mean much to me, this being my first day, but what I did not know was that Roy Chadwick was accompanied by the chief pilot of this South American airline and that his main interest was to see how the design of his 'Pee Tube' looked. I had my head down like everybody else and I did not notice that they had gone to my Section Leader and were now beside my board.

Chadwick: *"Where are the drawings for this Pee Tube ?"*

Me: *"I was just about to start, Sir".*

Chadwick looked at my board and the beautiful chain dot centre line. What went through his mind I cannot imagine.

*"Where are your sketches?"*

*"Er - I haven't done any Sir".*

This was too much for Chadwick. He more or less barged me off my stool and sat on it himself.

*"Right - scrap paper!"*

I did not have any scrap paper. Of course, I realised from that moment on, that this was an unforgivable sin. Someone produced some scrap paper but Chadwick was not going to let me off there.

*"What are you going to make it of?"*

he asked, looking directly at me. I felt struck dumb. Thankfully, my Section Leader came quickly to my rescue:

*"I'm thinking of 30 gauge stainless steel"* he said.

Chadwick wheeled round to me again.

*"Have we got any in the stores?"*

*"I don't know, sir...".*

*"Well get on to them, man!"*

By this time, some 150 pairs of eyes were on the shouting scenario around my board and on me in particular. I stood for a moment, I had no idea where the stores was.

*"Get on the phone to them! "*

All I can remember now was going to the internal phone behind the Section Leader's desk and fumbling with the 'phone directory. I cannot remember how it exactly ended.

I suppose Roy Chadwick led his visitor away making some suitable apology and assuring him that the job would be completed without delay by a competent draughtsman, and not that idiot they had just confronted!

I also imagine that as soon as his visitor had left he was on to the Chief Draughtsman asking him why he was employing hopeless characters like me in his Design Office. A case must have been made out that it was my first day there and that I was actually a pilot/engineer and in no way a designer. As you can imagine, I have never forgotten that day and my first meeting with the great Roy Chadwick. That day I learnt the hard way how a design task is approached and I have never been without scrap paper since.

So, having blown my chance to be an aircraft designer, fate took a hand soon afterwards when my Section Leader came to me and said:

*"Someone is needed to fly in a Tudor".*

I wasn't speechless this time!

*Roy Chadwick, Chief Designer of A.V. Roe & Co. Ltd. from August 1919 until his death in therTudor II accident on August 23rd 1947.*

## CHAPTER TWO

# Woodford Airfield, Cheshire

The airfield at Woodford has many a tale to tell and much aeronautical history to unfold. It lies on the Cheshire Plain some five miles south of Stockport and about five miles to the east of Manchester Airport.

*Woodford airfield in the 1970's - the original pre-war Lancashire Aero Club facilities and Avro Flight Test Sheds are in the top left of the picture. In the foreground is the huge Assembly Shed built in 1939. The main runway, (running left to right) was lengthened in stages out to the left to take the Vulcan.*

Before 1924 it was New Hall farm, but in that year it became the northern airfield for A.V.Roe & Co.Ltd. after the closure to any further flying at Alexandra Park in Manchester. (Hamble, near Southampton, was Avro's southern airfield from 1916 to 1928).

During the 1930's, aircraft such as the famous Avro 504 and Tutor were built at the Newton Heath factory on the other side of Manchester and assembled and flown at Woodford. Many of these aircraft were towed through the streets of Manchester on their own wheels to arrive at Woodford.

Beside Company activities, very good flying and club facilities were provided for amateur aviators, including the Lancashire Aero Club.

Despite the industrial depression in the 1930's, assembly and test flying of Avro aircraft continued steadily. However, the ominous threat of war with Germany and the resulting expansion of the Royal Air Force, changed life at Woodford into top gear .

The story of Avros and the intense activity at Woodford during the Second World War is a history in itself and the contribution made to the success of Bomber Command by the famous Lancaster is legendary. As the production build-rate accelerated,

*Five Avro Athenas in front of the old Lancashire Aero Club clubhouse and hangars at Woodford. The aircraft are (from left to right) production T.Mk.2s VR566 and 567; R-R Dart engined T. MK.1A VM129(with the original fin shape) and prototype T. Mk. 2s VW892 and 890.*

*Arthur Fearfield, an Avro Aircraft Inspector checks out the second prototype R-R Merlin -powered Avro Athena T.Mk.2 VW891, at Woodford. This was fitted with a larger fin and balanced rudder.*

*Above: Anson T.Mk.20 VS504 airborne over Woodford in February 1948 with Ken Cook at the controls.*
*Below: The Flight-Test Section in the early days (about 1949), with Peter Thorn (son of the late CTP 'Bill' Thorn) 3rd from the right back row, and the Author, 3rd from the left, front row.*

test flying was on seven days a week for more than three years such that the flight clearance rate rose from 20 aircraft a week to a maximum of 37. At one stage, when the airfield was fog bound for three days, there were over 80 Lancasters parked on the airfield and adjacent farmland. When the fog eventually lifted, Bill Thorn and Jimmy Orrell flew 29 flights in one day to ease the backlog.

Fortunately, Woodford was never subject to being bombed by the Germans. On one occasion two German bombers approached but turned away before reaching near enough to be in sight of such a prime target.

*The unpainted and un-registered prototype Tudor MK.I G-AGPF at Ringway just before it's maiden flight on 14th June 1945.*

After the war ended, Avros entered seriously into the business of civil aircraft. Using the basic design of the Lincoln bomber wing and four Rolls-Royce Merlin engines, several marks of the Tudor airliner were built. I joined the Flight Test Section about this time and I will always remember, as a young and inexperienced 24 year-old, being driven from the Design Office at the Chadderton factory down to Woodford to fly in a Tudor. In those days (1947), lunch was taken in the Clubhouse and it was a proper sit-down three course meal - none of your sandwiches eaten at the desk affair. The meal was served by Mr. Leyton, who with his wife, looked after the Clubhouse. The Chief Test Pilot, Bill Thorn, sat at the head of the table and I always

*A group picture taken by Avro Tudor II G-AGSU at the end of it's first flight on 10th March 1946. (L. to R.) Alf Sewart, Bill Thorn, Roy Chadwick, Teddy Fielding, Jimmy Orrell, Arthur Bowers and Jack Dobson.*

*Flight Test office - the large notice-boards on the right hand wall display the status of Avro Ashton testing and Shackleton M.R .1 development.*

*A Shackleton A.E.W.2 gets airborne on a test-flight.*

marvelled as to how Mr. Leyton knew exactly the right moment to come in from the kitchen to clear the plates ready for the next course. Eventually I discovered the answer - Bill had a push button on the floor near his right foot!

At that time there were six Test Pilots and three of us in the Technical Observers section. As young men we were very fortunate to be working in the front line of the Company's activities, so to speak, and to be a member of such an experienced flying team. The Managing Director, Roy Dobson (later Sir Roy) and the

Chief Designer, Roy Chadwick would visit us and sometimes to fly and so we had the added benefit of being in contact at first hand with the top management.

So in those days we were a technical team of three with a small wooden office and one cupboard which housed the only three pieces of test instrumentation that we possessed. But the section grew into a

*Oops! Who forgot to set the brakes then? Lancaster Mk.VII WU-03 (ex RT697 for the French Aeronavale experiences a 'coming together' with Anson T.21 WJ515 at Woodford in January 1952. It seems the Lancaster had the brakes incorrectly set and rolled backwards down the slope outside the Flight Sheds.*

*Morning break for the Flight Test Office staff at Woodford!*

*HRH The Duke of Edinburgh visits Woodford and meets the Test-Pilots on 8th November 1955. L. to R. Jimmy Orrell, (Chief Test-Pilot), Johnny Baker, Sqn. Ldr J. B. Wales, Jimmy Harrison, RAF Liason Officers Sqn Ldr C. C. 'Jock' Calder, (of 617 Sqn fame), Sqn. Ldr D. R. 'Podge' Howard and Flt. Lt. Yates. With HRH, and with his back to the camera is Roy Dobson.*

department and by the time the first Vulcan protoytpe arrived at Woodford, there were 120 staff with offices and test facilities to match.

Many new types of aircraft and the advent of turbine engines made the post war years one of the most interesting and exciting periods in aeronautical development for those of us fortunate enough to have been engaged in flight development work.

The name A.V.Roe, though never to be forgotten in aviation history, has passed on commercially. The design and manufacture of British and European Consortium aircraft is now in the hands of British Aerospace. Many of the Company's airfields and factories have been closed but we are pleased that Woodford, where we were lucky enough to spend most of our working lives, is still going strong. In fact I believe that Woodford is now the Avro International Aerospace of BAe and so that famous old name does live on and quite right too!

# Welcome Aboard!

On most airlines there is a pleasant welcome from one of the cabin staff when you first enter the aircraft you are going to travel in. Generally, this welcome and the cleanliness and preparation of the cabin gives a feeling of reassurance that the airline knows it's business and the flight will be comfortable and enjoyable. Few people if any, will give any thought as to what work went into the preparation of that aircraft to bring it up to that state of readiness.

If the particular type of aircraft has been in service with that airline for some time, the preparation work will have been, usually, smooth and routine. But what if it is the very first flight of a brand new type of aircraft and hence somewhat unfamiliar to the ground staff preparing it for flight ?

In this day and age there is advance training to make staff familiar with the systems and layout before they actually receive the new type, but it was not always that there was time for such rehearsals.

Late in the evening, on September 29th 1947, Tudor IV Reg.No. G-AHNK was being prepared for flight. It was being got ready, not at the airline's base at Heathrow Airport outside London, but at A. V. Roe's airfield at Woodford, Cheshire. This was the first aircraft going to be operated by British South American Airlines (BSAA) and was scheduled to take-off from Heathrow at noon the next day on a route-proving flight bound for South America with fare paying passengers on board. First however, there had to be a test flight before the Certificate of Airworthiness could be issued.

We took off from Woodford at around 8pm. The pilot was our Chief Test Pilot, Jimmy Orrell (See *"The Quiet Test Pilot"* by Peter V. Clegg). The second pilot was Air Vice Marshal D. C. T. 'Don' Bennett, the head of BSAA. In addition to myself, there was the flight engineer and two other technical observers, one of whom was the Air Registration Board representative. It was dark and in those days there were no runway or

*Avro Tudor IV G-AHNK 'Star Lion' of British South American Airways being flown by Avro's Chief Test Pilot Jimmy Orrell near Woodford on it's maiden flight on 29th September 1947.*

taxy lights at Woodford. Unperturbed, we aimed our take-off run to a point between two cars stationed at the far end of the runway with their headlights on.

The flight tests completed, we landed at Heathrow after about two hours. The next job that had to be done was to correct the results to Standard Atmospheric conditions and then write them up in an acceptable technical form. They were passed to the ARB representative who, after considering them, signed and released the C. of A. documents.

The aircraft was finally handed over to the airline around midnight, so leaving their staff less than twelve hours to stock and prepare a brand new aircraft type.

The next day, as the aircraft took off on the stroke of mid-day, I am sure none of the passengers would have any idea, or could imagine, what had gone on in the preceding twenty four hours.

# A Lightning Strike

After the Second World War, aircraft and aero-engine designers and manufacturers entered fully into the era of civil aircraft and turbine engines. At A.V.Roe we were producing Tudor airliners in various versions. At Rolls-Royce one of the turbine engines being built was the Nene. The two Companies came together to co-operate on an interesting project which replaced the Merlin engines on one of the Tudor aircraft with four Nenes. For those of us who flew in it, the sudden absence of the vibrations of piston engines and propellers was a revelation I will never forget.

At this time - and I am speaking of 1948/9 - there was considerable interest and excitement in the introduction of turbine engines.

At the main A. V. Roe factory at Chadderton, near Oldham, the staff seldom had the chance to see the fruits of their work flying and in response to a good suggestion, the Tudor VIII took off from Woodford and flew the twenty odd miles to circle low over the Chadderton factory. All the workers were outside to see it fly over but weather wise, it was not a very suitable day. There were low dark thundery clouds filling the sky and the air was pretty bumpy - not the best background to show off a new aircraft.

*The Tudor VIII VX195 with four Rolls-Royce Nene 5 turbojets. Converted from Tudor I second prototype G-AGST, it became the world's first four-jet airliner when it flew on September 6th 1948.*

*Operating a tailwheeled jet aircraft brought its share of problems as the above photograph of Jimmy Orrell getting airborne in VX195 at Farnborough shows. Because of the tailwheel, the then downward sloping exhausts scorched the tarmac!*

*The tailwheeled Tudor VII was re-designed into the Avro Ashton research aircraft, WB492 seen here.*

Lightning was flashing every now and then and one flash appeared to strike the aircraft. On landing back at Woodford, the lightning strike was reported to the ground staff in order that they could inspect the aircraft for any signs of it. Sure enough, along the skin of one of the tailplanes there was a line of dents and bumps and that portion of the skin was removed.

While this was going on the pilot was sitting in his office. There was a knock at the door and the foreman entered with a broad grin on his face

*" We have found a lump of lightning in the Tudor!"*

he said and he held out a block of metal which had inadvertently been left inside the tailplane during it's construction at Chadderton and was the cause of the dents as it bounced around inside the structure!

# Regrets and Relief

In my experience, most test flying is routine and uneventful and most of one's working time is on *terra firma*. By far the largest part of it is taken up with meetings, planning, analyses, report writing and other ground based jobs. Sadly there are bad days in test flying and even if one is not involved directly, there are both relief and regrets.

My first boss, the man who enabled me to escape out from the drawing office and enter into the world of development and test flying, was killed only a few months after I joined him in the Flight Test Section.

It was a Saturday morning and there were two Tudor aircraft to be flown that day. Sqn.Ldr. Wilson, head of our section, flew as second pilot in one of them and the only other member of our section, Stan Nicol, flew in the other Tudor. I was left working in our office which was located in the corner of the Experimental Hangar that still had it's glass roof blacked out from wartime days. There was no one in the hangar and all was absolutely still and quiet, being a Saturday morning.

All of a sudden I had quite a shock. The door of the office burst open and a very distraught lady shouted:

*" Is anybody left, is anybody left?"*

*The ill-fated Tudor II G-AGSU, with modified fin and rudder seen on the ground at Woodford. At this time (1946-7) it was the largest civil aircraft produced in Great Britain and some 79 had been ordered. In the event, only 4 were produced.*

*Avro Sales Brochure for the Tudor II, featuring G-AGSU on an earlier flight with its original tailplane. Ironically, on the day it crashed it was taking off in the opposite direction, and was at this exact spot and height when the crossed aileron controls began to exert a continuous bank to starboard, resulting in a side-slip nose-down into a pond beside the runway.*

THE AVRO

# TUDOR II

A·V·ROE & CO LIMITED  •  MANCHESTER  ENGLAND

I was quite startled after the absolute silence of the previous half an hour.

*"There is only me here I'm afraid, who were you looking for?"*

She did not answer but turned and ran out of the hangar. It seemed very unusual behaviour and I decided to go out and see where she had gone. I did know that she was not an intruder as I knew her slightly as the wife of one of our flight engineers. Outside the hangar and by the time that I was halfway across the tarmac apron, I realised from the fire engine and ambulance activity that something was seriously wrong.

I ran across the airfield and came upon the scene. Tudor II, G-AGSU had crashed nose down into a deep pond at the side of the runway. The investigation that eventually followed found that the aileron controls had been crossed.

Fate decreed that day, that some would fly in the G-AGSU and others who had intended to or who might well have done, did not.

Our Managing Director, Roy Dobson was on board but was called away to take a phone call...

Test pilot Jimmy Orrell had tossed a coin with Bill Thorn on the day before as to which of them should fly the aircraft - Bill lost the toss...

Because it was a Saturday, our Chief Designer Roy Chadwick decided to join the flight...

Of our section, Wilson as head elected to go in 'SU, Stan Nicol to fly in the other Tudor and yours truly to guard the office...

...Fate played a strong hand that day.

Many years later - after completing some performance trials in Cyprus - we were returning to the UK in the trials aircraft, an Avro 748. We had refuelled at Rome and were on the final leg to Manchester. It was quite

*The 2nd prototype Avro 748 and company development aircraft, G-ARAY.*

dark by the time we were over the Alps and round about this point in the flight, the pilot called me up to the cockpit and pointed to the power instrument for the Starboard engine. This was indicating significantly less power than the Port engine although both were at the same throttle setting. Even more worrying was the fact that the power was continuing to reduce. After about another half an hour the power on the Starboard engine was so low that the engine had to be shut down and the propeller feathered.

By this time the power of the Port engine had also started to fall and was down to about half of normal. We were over the English Channel when the pilot decided to send out MayDay calls. Fortunately the calls were answered by Lydd airport and by skilful piloting, we virtually glided down and landed safely from the first approach. This was just as well, as there was no possibility of a go around to make a second attempt on a quarter of an engine.

The airport staff were very happy to receive us except for one customs officer who had to be called in from home, the airport having long since been closed to

*The entire tropical trials team - Test-Pilots, Flight Test staff and Ground Engineers in Cyprus, along with 748 G-ARAY.*

normal traffic for the night. We booked in at a local hotel for the night and spent the evening in the bar. There was a lot of mind searching discussion on the incident but the only possible reason for the loss of engine power that we could come up with, was that the fuel we took on board at Rome might have been contaminated in some way. After all we argued, we had no sign of the trouble on the flight from Cyprus to Rome. On the strength of this suspicion we sent a telex to Rome Airport to warn them of this possibility.

The next day, with the help of some engineers who had flown down from Manchester, the reason for the problem soon became clear - I could have kicked myself. Being a test aircraft, we were flying with special test fuel flowmeters fitted which incorporated felt type filters. Flying for several hours in the colder high altitude air over Europe had caused tiny ice particles to build up on the filters and increasingly restrict the flow of fuel. These flowmeter systems include an electrical switch to enable the filters to be bypassed for take-off. So all we needed to have done on the flight from Rome

*Jimmy Harrison, Chief Test-Pilot leads Colin Allen, Bob Dixon-Stubbs and the remainder of the staff aboard G-ARAY for another flight -test from Cyprus.*

*Sharing an early morning joke at Nicosia Airport during the testing of 748 G-ARAY is the author, Fred Hill and Bob Lawson.*

was to have operated these switches. Making sure these switches are in the 'bypass' position before a take-off is one of the first things a technical flight observer is taught. It was because we were cruising at 18000ft. that the reason for what was causing the fuel starvation did not occur to me. Nevertheless, ever since, I have never forgiven myself and I will always regret that, on that day, I failed technically.

There was an embarrassing postscript to this story. After some modifications to the aircraft, we flew back out to Cyprus again via a refuel stop at Rome. As we taxied up to the refuelling area at Rome airport, one of the fuel company's staff ran out to meet us holding aloft a glass beaker of fuel. His message was obvious:

*"No dirt in our fuel, no dirt in our fuel, how dare you!"*

...or words to that effect in Italian!

In November 1956 we were carrying out stall warning tests on the Shackleton M.R. 3 prototype, WR970. This

aircraft did not have an adequate natural warning of the approaching stall condition eg. buffeting or shaking. Cyril Bethwaite, who was Head of the Flight Research and Development Dept. at the time, decided to introduce an artficial stall warning system. This consisted of a small metal flag mounted on the upper surface of the wing at a point of critical airflow in the near stalled flight condition. When the airflow over the wing was normal it held the flag in the trailed position such that the electrical transmitter mounted inside the wing and in which the flag mast was mounted, passed no signal to the oral warning device in the cockpit.

*Avro Shackleton M.R. 3.WR972, with airborne lifetboat fitted, is seen on the ground at Woodford.*

When the aircraft was flown slow enough for the airflow over the wing to become turbulent and locally reversed, the spring loaded flag would turn 90deg. causing a switch in the transmitter to send a warning to the pilot that incidence must be decreased and airspeed increased, if a full stall was to be avoided.

This device worked very well but was a little marginal with the aircraft flying in certain configurations of radome and flaps. Nevertheless, it was considered worthy of an assessment by the Aircraft and Armament Experimental Establishment at Boscombe Down in Wiltshire. So John Baker and myself flew the aircraft down to Boscombe where we carried out nine flights over the next ten days with an A.&A.E.E. pilot to carry out the assessment. The weather was generally good and relatively cloud free, which was fortunate, because the tests involved going right through to the full stall and on this particular aircraft the starboard wing stalled first resulting in a very sharp wing drop to an angle of bank of 90 degrees and sometimes more. It was therefore crucial that the tests were started at a safe height and with no cloud immediately below, so that as the aircraft descended, the pilot did not become disorientated and could recover the aircraft from the stalled condition.

The result of the assessment was that the stall warning system was not acceptable in certain aircraft configurations and so we flew the aircraft back to Woodford to re-consider the problem. After two days and some changes to the warning system the flight tests were resumed from Woodford. A different test crew flew in the aircraft, and for reasons I cannot recall now, neither John Baker or myself were on board.

The aircraft crashed over the Derbyshire hills and sadly all lives were lost.

It was not the intention of this collection of tales to dwell on the tragedies but as I sit here looking back 30 or 40 years, I realise what a strong hand fate played in our lives. At the age of 15 years I had no idea as to how I might earn my living. Just a chance remark by a friend found me in the RAF and from then on my life was wrapped up in aircraft.

# Triangles at Boscombe Down

With the advent of turbine aero engines, Vickers, Handley Page and A.V.Roe were called upon to design and build new bomber aircraft utilising these new engines. Avros decided to design their aircraft with a delta wing planform and without a tailplane. This was a major step forward in aircraft design, because at that time, little was known about the characteristics of such a wing planform other than aerodynamic theory and wind tunnel test data.

To acquire more knowledge by actually flying the proposed wing planform, ahead of the final design of the wings for what would become the Vulcan bomber, several third scale research aircraft were built and test flown.

For a number of reasons the third of these research aircraft, designated Type707A, was flown at the Aircraft and Armament Experimental

*The author (far right) enjoying a party at the Officers Mess, Boscombe Down. Sqn. Ldr. 'Bill' Sheehan, (far left) and Wg. Cdr. MacDonald (third from right) were both Shackleton pilots.*

*The new shape in the sky - the Avro 707 delta research aircraft. This was a third-scale flying example, and forerunner to the mighty Vulcan. Sadly, the first machine, VX784 crashed less than a month after it had first flown, killing Avro's Deputy Chief Test Pilot Sqn. Ldr. Eric 'Ricky' Esler...*

Establishment at Boscombe Down, near Salisbury. A small technical team from Avros moved to B.D. to support the flying programme and analyse the test data.

We were about ten in number and the A.&A.E.E. gave us hangar space with an adjoining office for the ground engineers. The technical staff worked in an RAF hut complete with a very welcome coke burning stove in the middle. Living accomodation was at the Cathedral Hotel in Salisbury and the Company had provided the team with a Ford van to get us there and back and for various tasks around the airfield including the fetching of coke for the office stove.

These were quite exciting times and some valuable test results and conclusions were passed back to the Design Office at the main Avro factory at Chadderton. We felt like technical pioneers and many VIP 's visited B.D. to see how the aircraft flew  and to discuss progress.

When not flying, our aircraft was housed in what was designated the 'Naval' hangar, which for most of the time was fairly empty. Walking through this hangar one day, on return from lunch, I came upon  a crowd of twenty or so Naval personnel gathered together in the centre of the hangar. Being nosey, I went over to see what was happening. It appeared, that one man was holding a small sparrow in his hand. The poor creature

had broken one leg and it was hanging loosely down and was obviously of no further use to it. An argument was going on as to whether the bird should be released to fend for itself as best it could, or whether it would be more humane to put it out of its misery. It was about a fifty-fifty argument with the bird having no say in its future at all.

Then up strode the man to settle the problem. A huge hunk of a man, well over six feet and who in this day and age would have been a likely candidate for the television programme 'Gladiators' . He pushed his way through to the centre of the crowd and quickly weighed up the situation.

*... the later 707B, VX790 was first flown by 'Roly" Falk at Boscombe Down on September 16th 1950. It is shown here with the first of two different types of dorsal engine intakes that were fitted. The 707B was a low-speed test vehicle for the Vulcan.*

*"Don't piss about! Give it to me".*

He took the poor little sparrow into his huge matalot hands and immediately wrenched its neck to finish it off. But the sparrow still showed signs of life and so the huge hands wrenched the neck again, this time a little harder.

*" There you are, that's the best thing for it ".*

He opened his hands to let the dead bird fall to the ground and it flew happily away.

34

I thought afterwards, perhaps that is why there were never many aircraft in that particular hangar!

Hangars, where aircraft are housed and worked on, are potentially dangerous places and fire and other precautionary rules have to be strictly adhered to.

Being a research aircraft, our 707A had special test instrumentation installed. The panel on which these instruments were mounted was photographed during flight tests by a 35mm. ciné camera ( very old hat these days, but this was back in the 1950's ). In the checking of this set up, invariably quite a lot of film was discarded from time to time. Our instrument engineer had piled this unwanted film into a dustbin in the office, which was situated at one side of the hangar. The film was not wound up  but just lay loosely in the open dustbin.

One day, when all had been going well, the office had a visit from the A.&A.E.E. Chief Fire Officer. After introducing himself, he gazed around the office until his eyes fell upon the dustbin, now quite full of loose film. I cannot remember exactly what he said at the time, but I do remember that he left us in no doubt that he was horrified and that we were endangering the whole hangar with such a potential fire hazard. But our instrument engineer soon put his  mind at rest.

*"Don't worry, it is non-inflammable film - watch"*

With that he struck a match and threw it into the dustbin. The film burst into flames, some nearly reaching the office ceiling.
Of course, our faces were red but the fire officer's went purple and when he had recovered, he had a lot more to say about the matter. They say - Never try to prove a negative! ( Excuse the pun ).

## CHAPTER SEVEN

# How high
# the Vulcan?

Early in the flight development stage of a new aircraft the Static Pressure Error must be determined experimentally in flight over the full range of altitude and Mach number.

Static pressure error may be defined as: 'Indicated height' minus 'Pressure height' where 'Pressure height' is the height corresponding to the true atmospheric 'Still Air Pressure' in the vicinity of the aircraft and 'Indicated height' is the height registered on the pilot's altimeter.

To determine the static pressure error, the pressure height is measured and compared with the indicated height recorded at the same instant. One method of measuring pressure height uses a trailing static head suspended 50ft or more below the aircraft, where the

*Fig 1: the trailing static is just visible below and behind the Vulcan in this somewhat poor quality test photograph.*

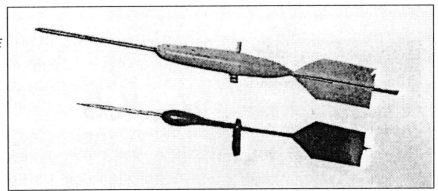

*Fig.2: The standard (lower) and modified RAE low speed trailing static heads.*

air is undisturbed (as shown in Figure1). Two types of head are shown in Figure 2. The lower one shown is the standard type of head for use in low speed and Mach number flight. The upper one was the modified head for use with the Vulcan, and as can be seen from the results of wind tunnel tests shown in Figure 3, it incurred smaller basic pressure errors in the higher Mach number range. In the event however, this head was not used at Mach numbers above 0.65 during the trials to measure the pressure errors of the Vulcan

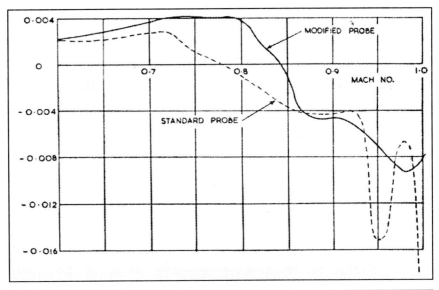

*Figure 3: The calibration of trailing static heads in the Avro transonic wind tunnel.*

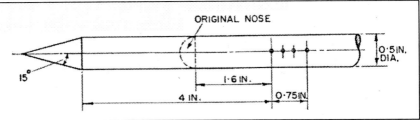

*Figure 4: the modified static head showing the holes.*

B.Mk.2 static vent system. The static holes in the nose of the head are shown in Fig.4. The holes are connected by a flexible pressure tube to an aneroid barometer in the aircraft.

Trailing static heads work very well in providing a source of true atmospheric Still Air Pressure as a datum but they have a Mach number operating range well below that of the Vulcan. At the higher speeds they will break away and at the higher Mach numbers, the drag rise can promote an unstable phugoid and the head will no longer measure true static pressure. (It is believed that a winged T/S head has been developed at the RAE for use at higher Mach numbers but at the time of the tests described here, it had not been cleared to meet the full Mach number range of the Vulcan).

To ensure that pressure errors (P/E's) of the aircraft's static source for normal operating flight and for flight test analyses are well established, it is essential to measure P/E's to the practical limit of the aircraft's flight conditions. This entails flight at near maximum altitude if measurements are to be made in steady 1 'g' flight; test points at the higher values of lift coefficient should not be established solely in 'pull-outs' or 'turning flight'.

To obtain the most accurate results for the Vulcan B. Mk.2 a completely new test method was conceived and put into use by Avro at Woodford. It was adopted because it promised to fulfil the following requirements:-

(a) All instrumentation and equipment required is contained in the test aircraft, so giving complete independence of other aircraft or ground stations.

(b) Measurements can be made over a wide range of height, lift coefficient and Mach number in level and curved flight paths.

(c) There is no restriction on geographical location for flight tests, other than being over the sea.

(d)  Only average test flying weather is necessary.

(e)  Complete technical and administrative control can be maintained over all aspects of the test, instrument calibrations and analyses.

(f)  Reasonably quick execution is possible, so enabling a worthwhile number of test points to be obtained.

(g)  The method is independent of wind speed and air temperature.

The method was called the "Radio Altimeter-Trailing Static Method of Measuring Aircraft Static Pressure Errors".

**Description of the method:**
The tape height of a strip of airspace was calibrated in terms of pressure height by flying a series of low speed calibration runs with a trailing static head suspended below the aircraft and a high-altitude radio altimeter operating in the aircraft. A continuous record was made of the tape height given by the radio altimeter and the pressure height given by a sensitive aneroid connected to the trailing static.

A minimum of three calibration runs were carried out, one at the nominal test height and two at other heights at suitable levels above and below the test height. Airspeed was held constant for steady level flight, consistent with satisfactory behaviour of the trailing static head being towed below and behind the aircraft.

On completion of the three calibration runs, the trailing static head was wound in and the P.E. test runs were made within the calibrated air space at the required test speeds. Continuous recordings were made of the radio altimeter height and of the pressure readings of the aircraft's static pressure system for which the position error was required to be known.

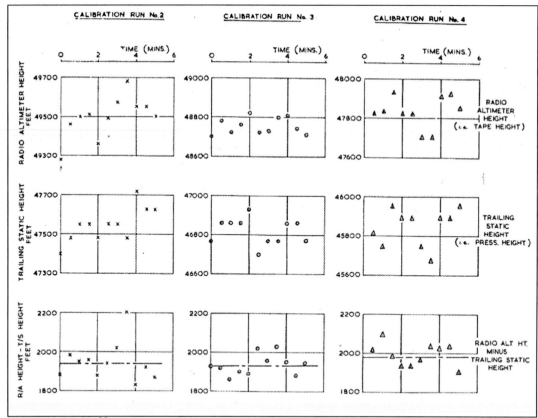

The differences between the pressure heights measured by the trailing static head and the radio altimeter 'tape' height obtained during the calibration runs, were plotted and are shown in Figs. 5 and 6. The mean difference at each height was applied to the 'tape' heights measured by the radio altimeter in the subsequent P.E. runs, so giving a true pressure height for each test run.

After the Position Error of the the aircraft's static pressure system has been measured at high altitude they and the results at lower altitudes using more standard methods are converted into the equivalents in terms of speed, height and Mach Number for inclusion in the aircraft's Flight Manual.

*Figures 5 (above) and 6 (opposite) : A pair of charts detailing calibration of the radio altimeter in terms of pressure height.*

(The above is a shortened description of this particular test method developed and satisfactorily used by Avros in 1962. The full description of the method was printed in the Journal of the Royal Aeronautical Society, Volume 67, No. 627 dated March 1963.)

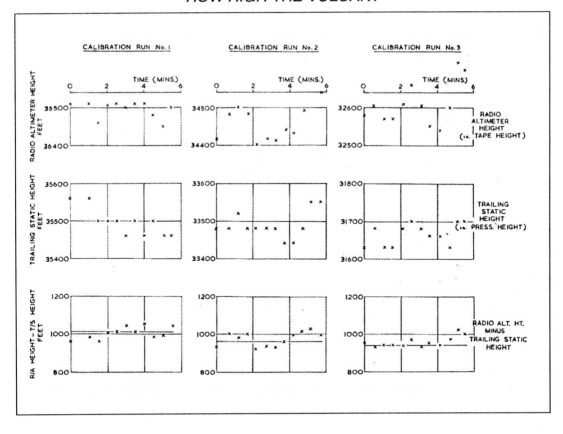

## CHAPTER EIGHT

# Vulcan XA891

One day in July 1959, Jimmy Harrison lifted XA 891 clear of the Woodford runway, retracted the wheels and had only just settled into the climb when all the main electrical generation system failed. Some of the flying controls stopped working and the 112 volt back-up battery showed less than 70 volts! All attempts to bring the system back on line were unsuccessful.

To turn and land at Woodford would have meant approaching over the village of Poynton. This was considered far too risky and so climbing flight was continued Eastwards. Contact with design engineers on the ground did not bring any solutions to the problem and it was realised that shortly, power to the flying controls would be lost altogether! Two of the rear crew members who could contribute nothing more to the situation were invited to leave the aircraft and did so, Bob Pogson, manning the electrics, remaining.

Some way short of the Humber Estuary, the flying powered controls stopped completely and the aircraft

*Vulcan XA891 before the accident. This was the third production B.1, but it had been retrofitted with the B.2 type cranked leading edge and was flying with a mixture of Bristol Olympus 200 engines.*

Avro's Chief Test Pilot and flying XA891 on this occasion:- Sqn. Ldr James Gordon 'Jimmy' Harrison OBE, AFC, C. Eng.

was no longer controllable. The Mayday call went out followed by "Abandon aircraft". All the crew landed safely on *terra firma* while poor old 891, left to it's own devices, landed on farmland north of the Humber. By any standards, it was not a good landing and the result can be seen in the 'after' photograph.

Jim Harrison picked himself up and after disengaging himself from his parachute made his way to the road nearby. Of course Jim was still in his high altitude flying suit and helmet, so when the first car came along occupied by an elderly lady and gentleman, they stopped to assist what they probably thought was a stranded motor cyclist.

*" Can we give you a lift somewhere?"*

A kind offer but all Jim wanted was a smoke.

*" Do you by any chance have a cigarette you could let me have?"*

*" No! We do not smoke and in any case, I do not allow smoking in this car."* retorted the woman.

*" Well thank you anyway."* said Jim.

*" We are on our way to do some shopping."* the man said and off he drove.

A van came along shortly after. *" Sorry mate, don't smoke".*

Then the police patrol car drew up. Jim got into the back seat confident that the law would not let him down but the two upright custodians in the front seats declared: *" We never touch tobacco, Sir"*

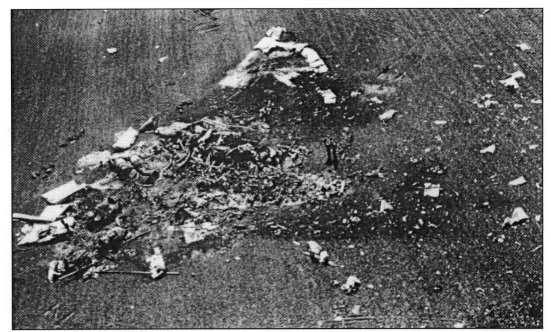

*The remains of XA891 in a field near High Hunsley where it crashed on Friday 24th July 1959, close to the Humber Esturary. The two people standing in the middle of the wreckage provide scale to the picture.*

As desperate as he was, Jim had to resign himself to the fact that there would be no smoking in the helicopter which arrived shortly after to pick him up. As they approached RAF Leconfield Jim was consoling himself with the thought that it would only be minutes before the hospitality of the RAF would have him in the mess with a glass in one hand and a beautiful cigarette in the other.

On landing, the hand of welcome was certainly extended and waiting, but then it directed Jim, not to the mess but to the sick bay. There the MO was waiting:

*" Here you are, take one of these now and one every 15minutes, preferably with a whisky!"*

He handed Jim a full packet of Players cigarettes.

When all the crew finally assembled in the mess at RAF Leconfield for a very necessary 'pick you up', Jimmy Harrison asked Bob Pogson as to how he had got on. "Well," said Bob, "I nearly went on the fastest slimming course in history". Apparently, in the heat of the moment he had forgotten to select the door handle to 'Emergency' and as a result, the door started to close as Bob slid down it.

*The crew of the ill-fated XA891 after the crash, and about to board Anson G-AGPG back to Woodford. L.to R.; Dick Proudlove, Bob Pogson, Phillip Christy (Bristol Engine rep), Ted Hartley, Jim Harrison.*

Thanks be to God, all the crew had the chance to get out of this one safely, which is more than can be said for the rear crew of another Vulcan which, on returning from Australia crashed in bad weather just short of the runway at Heathrow.

# Lucy's Treat!

To develop the Vulcan, large panels of extra instruments were installed in the prototype test aircraft and photographed with 35mm. ciné cameras during flight (very 'old-hat' these days). The cameras had magazines each holding 400ft. of film and after the film was developed it was projected on to special screens so that instrument readings required could be logged down for the technicians to analyse. The reading, logging and calibrating was an enormous task and the Film Reading Section, as it was known, was staffed by some twenty or so ladies led by a 'mother to them all' - Lucy.

## LUCY'S TREAT

To be a test pilot is many boy's dream
For in aircraft circles they're really the cream
The thrills and the glamour, the true chosen few,
Thought of as legends and sometimes it's true.
But a test pilot's life is not always a thrill,

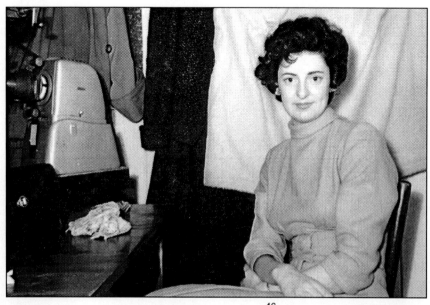

*Some of Lucy's film stars! Sitting beside a projector is Valerie Twiggs...*

## LUCY'S TREAT!

It's fraught with real danger and fear of a spill.
Misread your knots and you're really in trouble,
Thoughts of that great plane reduced to a rubble.
So many dials on the panel to read,
All are important, besides height and speed,
A.S.I, 'G' , R.P.M.'S, T.G.T.
A pilot daren't lose concentration, not he.
Except, when at AVROS one bright sunny day
Lucy applied for a ride in G-ARAY!
The girls were all thrilled, and answered the call,
But the pilot's misgivings never counted at all!
Six, all in mini-skirts, what consternation!
Pity the pilots, what price concentration?
Up through the clouds and into the blue,
The girls were relaxed and so were the crew,
When an engine snag caused them to colour with shame,
Though they were convinced that the legs were to blame.
They put down in Devon, a call to their wives,
The girls? They were having the time of their lives!
Back at AVROS meanwhile, suspicions abounded,
To hear that their pilots were finally "grounded".
Three ladies apiece and a night on the town?
Why, everyone knows G-ARAY never breaks down!
Of course that was only a joke, one supposes,
I told you a pilot's life isn't all roses!

*... with Vi Massey, Diane Milner and Anne Hardy.*

# The Vulcan meets the Barrier

The specification for the 'V' bombers drawn up after the Second World War called for a maximum Mach number of 0.95. This limit was somewhat arbitary in the case of the Vulcan, as at that time, there was little knowledge of the effects on delta wings of Mach number compressibility. As is well known, several delta winged aircraft to one third scale of the Vulcan design were built and designated Avro 707, Avro 707B (low speed), 707A (high speed), and 707C (two seater version). Sadly, the first 707 built crashed near Blackbushe just before the 1949 SBAC show, killing Avro's Deputy Chief Test Pilot Eric (Red) Esler.

Most of the investigative flying on the 707A was carried out by 'Roly' Falk with the aircraft based at the A.&A.E.E., Boscombe Down. The author was lucky enough to be a member of the supporting team under the leadership of Z. Olenski, a Polish ex- Battle of Britain pilot. The development flying on the 707A provided some very useful learning on the aerodynamic effects on the delta wing at high speed and high sub-sonic Mach number.

Although of course the Vulcan was never designed or intended to be a supersonic aircraft, the Mk.1 version had a pitot static system that had a 'position error' such that the Machmeter overread. When flying

The second prototype Vulcan B.1 VX777 after being fitted with a new, cranked wing leading edge is seen here establishing the setting for the probe at the tail end that operated a 'nose too high' warning light in the cockpit.

*Avro Test Pilots Tony Blackman and Jimmy Harrison in front of a Vulcan. Tony succeeded Jimmy as Chief test Pilot when the latter gave up flying in 1969 (becoming Product Support Manager at Woodford for the next 14 years).*

at what was probably a true Mach number of 0.94/0.93, the Machmeter in the cockpit would indicate 1.01. An amusing tale was brought back after the Vulcan had taken part in the bombing competion in the USA. An American general had been given a flight in the second pilot's seat during which the high Mach number handling had been demonstrated to him. After the flight he was able to tell his fellow officers that *"we went supersonic! "*

The pitot static system on the Mk.2 Vulcan was changed to one that gave an indication much nearer to the true value. During test flights to demonstrate that the aircraft could satisfactorily achieve flight at the specification Mach number of 0.95, it was found that the elevon-up angles required to recover from diving flight increased markedly beyond 0.93. As a result, dives to Mach numbers above this value were made in very cautious small steps. At an indicated 0.955, Jimmy Harrison had to call out to the second pilot to pull on the control column as it had become more than he alone could manage to pull out of the dive. This situation of course, was far from satisfactory and meant that in this configuration the aircraft would have to go to the A.&A.E.E., Boscombe Down for acceptance trials with a recommended limit of no more than 0.94.

The brains of the design and the flight development staff had been working hard for some days on possible modifications to the control system to alleviate the problem. The elevons were fully power controlled with an artificial feel unit providing increasing stick force to the pilot's control column as speed and elevon angle were increased. Fig.1 shows the arrangment diagrammatically.

The first obvious thoughts on a solution were to feed Mach number signals into the artificial feel unit so as to move the fulcrum in order that the input force had a better mechanical advantage over the main spring as Mach number increased through the critical range. This idea was all right in principal but would have been difficult to engineer as the fulcrum was already positioned by a signal from airspeed data. In addition, a more conservative limit would have had to have been

*Figure 1: The spring strut modification.*

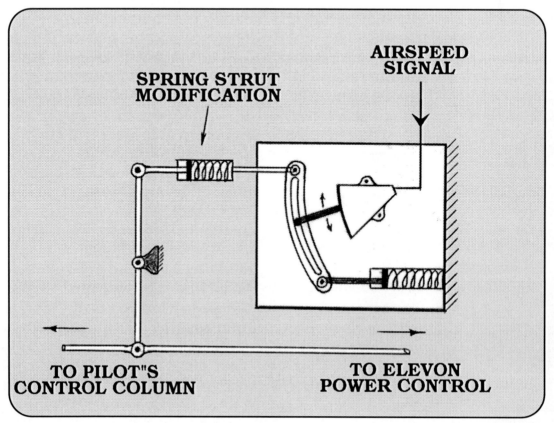

applied to the Maximum Mach number clearance figure to allow for production tolerances in instrument and position error. Pondering on these aspects, one of the Flight Development staff came up with a much more positive, cheaper and easy to engineer solution and that was to introduce a pre-loaded spring strut between the pilot's input and the artificial feel unit. The spring strut would have a break-out force equivalent to a safe but manageable force at the pilot's control column and a figure of 150lb. was decided on.

Fortunately, the final push-pull rod into the artificial feel unit was of a suitable length to accomodate such a strut (see Fig. 1).

The test pilots, while acknowledging this idea would be ideal if it worked, were a little concerned that in pulling out of a dive they would feel a discontinuity, or a step, in the pull force they were applying. Fortunately again, we had set up in one of the hangars, a complete Vulcan flying control rig. The spring strut was made and installed in the elevon circuit in the rig and the pilots were invited to come and pull with a gradually increasing force up to the maximum they could manage. None of them could detect any sudden step or discontinuity in the pull force v. elevon deflection slope, though there would of course, have been a reduced slope after the break-out of the spring strut had been exceeded.

So hats having been retrieved after throwing them in the air so to speak, the spring strut modification was issued officially by the Design Office and incorporated in all Vulcans including of course, the one sent to Boscombe Down for acceptance clearance on behalf of the Ministry of Defence.

This of course was an example of a problem very satisfactorily solved. Cheap to produce and install, the minimum of ground and flight testing and reliable and consistent in operation. Needless to say, not all our problems were solved as easily as this one was.

# Trials in Africa

Soon after we started measuring the performance of our Twin Dart 748 Series 2 in the UK, we needed to consider where we would have to take it for hot weather and high airfield testing. Africa offered the best choices and so I was sent off there to look at a short list of airfields. The list of possibles included Kano, Johannesburg, Harare, Nairobi and Khartoum and so my route was down the West side of the African continent and back up the East side. The journey would be nearly 12000 miles altogether and I planned my flights such that I could have two days at each place and the flying in between would, as far as possible, be at night. On the basis that I should have a good night's rest in flight, I was allowed to travel First Class.

For such a complex itinerary it all went very well with one exception, which I shall come to later. All flights were on time and there was always plenty of

*All aspects of flight performance had to be checked. Low altitude performance of the first protottpe 748 G-APZV is seen being sampled here at Woodford.*

*Senior Avro designers and technicians - including Dick Connor, Z. Olenski, Roy Ewans and John Scott-Wilson listen to Jimmy Harrison after the first flight of the Avro 748.*

room. In fact, on the leg down to Jo'burg, there was only one other passenger in the First Class cabin and he turned out to be the Dutch Ambassador to South Africa. To pass the time the cabin crew ran a competition with a bottle of champagne as the prize. All one had to do was to estimate how many hours that particular aircraft we were flying in had clocked up in it's life, after they had told us the date that it first flew. Well, I thought, here am I, an aircraft engineer and only one other person to compete against - that bottle of 'champers' is as good as mine! I did some careful calculations and came up with a number that I was sure must be close to the right answer. It wasn't, the bottle went to the Ambassador! What they had not told me was that this particular aircraft had been grounded for two years for modifications. I suspect that the Ambassador, who was a regular passenger on this flight, knew about this beforehand!

I had written to the manager at each of the airfields that I wished to survey, asking for a meeting. The managers at Johannesburg, Harare and Nairobi all saw me and were very cooperative. When I reached Khartoum however, I got a bit of a shock. The last time I had been there was in 1950, when I was part of the team carrying out tropical trials on the Shackleton

M.R.1. The Grand Hotel, although old fashioned, had been quite adequate in the past, but now was so run down that I could hardly believe it. The food was virtually inedible and during the evening meal, three Sudanese waiters started to fight each other in the middle of the dining room. I could only think that one had threatened the other two with ordering them to eat some of the food!

The next day I turned up at the airport manager's office for the pre-arranged meeting. There was a soldier outside his door and when I told him my business he went inside the office and I waited outside. He was out again in a minute and said *"The manager no see you"*.

I was a bit taken aback and I explained to the soldier that I had come all the way from England to talk to the manager . I showed him the correspondence and asked him to go in again and explain this to the manager. He went in, not only to get another refusal

*Aircraft Herald, Kano Airport, Nigeria, guarding the second prototype G-ARAY.*

*This 748 first flew with Rolls-Royce R.Da.6 engines, but was later re-fitted with more powerful R.Da.7 Darts in November 1961, becoming the Series 2 prototype.*

but I could hear him getting an almighty bollocking. Poor chap, he came out very upset. I apologised to him as best I could, didn't risk doling out any 'baksheesh' and mentally crossed Khartoum off my list. Back at Woodford once again, I wrote up a report on the trip and from this we chose Nairobi and Kano.

The tests at Nairobi went very well and the weather was typical for Kenya, blue skies with small puffy white clouds starting to form mid-afternoon. Being at 5000 feet, the temperature was in the mid-eighties and feeling very comfortable, so much so, that one can get caught out with respect to sunburn. Parked next to our aircraft on the apron was a Vulcan flown in from Aden. RAF fitters were working on top of the wings and, coming from Aden, they already had an established suntan. They worked in just their khaki shorts and no doubt felt very comfortable but after two days they had to report sick with a serious case of sunburn.

The ten days at Nairobi were very pleasant.We stayed at the New Stanley Hotel and one evening when we happened to be in the foyer, who should stroll in but

*748 G-ARAY in the company of an RAF Vulcan (thought to be XH480, a B.Mk.la of 101 Sqn) and Ethiopian Airlines DC-6B ET-T-27 at a rather damp Nairobi Airport*

Jomo Kenyatta. He had a chat with the boy on the door but didn't seem to notice us. Some of us managed a game of golf and soon learned to keep out of that rough - worse than heather! A trip up to Nyeri and an invite to a coffee plantation all made our stay most enjoyable.

Flying our aircraft from Nairobi across to Kano in Northern Nigeria via the Sudan and Chad was also interesting. We first flew from Nairobi to Juba situated in the very southern part of the Sudan and close to the White Nile. In the days of flying the East African "Empire" route by Imperial Airways, Juba was an overnight refuelling stop. Jimmy Orrell, who was a Captain often flying this route recalls:

*"Sleeping in a small straw Kraal at Juba with two wire-framed doors, and leopards prowling round after the hotel dogs, was apt to be interesting!"*

From Juba we flew across south-west Sudan for a refuel and night stop at El-Geneina situated just inside the western border of the Sudan. It is hard to describe the first impression one got as we stepped out of the aircraft here, though I certainly felt that I had really arrived in 'the middle of nowhere'. Flat desert with not a single relieving feature to leave a pleasant picture in one's memory and I only recall what happened there.

By good forward arrangements with Shell, the fuel we needed was there and waiting. The novel aspect of this was that we had to form a chain-gang line to pass cans along and up to those on the wing to pour the fuel in by hand. While this was going on, a large dirty old American car pulled up at one of the tin huts and a man carrying a couple of bottles got out and went inside. We learnt that this was the airport manager and that we would probably not see him again. We didn't and I thought how strange that he could not find the time to just come over and say 'hallo'. We did not see the town itself and stayed the night at the airport. The accomodation was in a long block of single rooms, as basic as could be but adequate, and before we retired they provided us with a surprisingly good meal.

*Local onlookers watch us get 748 G-ARAY ready for another test flight from Kano Airport.*

The next morning we were relieved to get airborne and thankful that nothing had gone wrong, though I must say it was an experience to have visited such a desolate place. I would not bother looking for it in books listing recommended bed and breakfast places in Africa!

We set course due West for Kano some 800 miles distant, first crossing the southern part of Chad. It was desert all the way as far as the eye could see, even from a height of 15,000ft. and yet it was sort of challenging to try and spot the odd feature, a small village or town, a range of rocky hills or even just a bit of scrubland.The one big feature that we did see in the distance, was Lake Chad.

Kano is quite a large Moslem city in the north of Nigeria. We stayed at the Central Hotel and hired cars to get us to and from the airport daily for our work. The drive was along a tarmac road with no pavements, just desert either side. Driving to work one morning, a man on a bike with a pile of folded sacks on his handlebars, rode off the sand on to the road and straight into my rear wheel. It was obviously not my fault but he fell off, grazed his knee and slightly bent his front wheel. We stopped and got out to see him sitting on the sand beside his bicycle looking rather dazed. A crowd of locals gathered around in no time and just stood staring at the scene. He did not speak any English and so having ascertained that he was not seriously hurt, I pulled out the equivalent of a ten pound note at the sight of which, the crowd gasped. I offered it to the man at the same time pointing to his damaged bike. He took the money and we were just walking back to our car thinking that was the end of that, when up rode a policeman. He could speak some English and after hearing the crowds version of events, he said that I should go with him to the police station. I said:

*"No, I have to go to work!"*

...but I agreed to meet him at the hotel at 4pm. that afternoon. Knowing that he had made a note of our car number I did turn up and he was there. I did not recognise him at first as he was no longer in his uniform. He handed me a scruffy piece of yellow paper on which was written:

*"The inspector recommends the arrest of the driver".*

I left him wondering what to do next and went away to ring the Police Commissioner who I knew was British. It turned out that he was away for a couple of days and in the meantime, Bob Lawson, who looked after admin. in our team, bought a ticket and booked me on a flight home, just in case. Luckily the ticket wasn't needed, the Commissioner arrived back and quickly sorted out what he afterwards told me was a case of trying blackmail on me.

So we finished our flight trials in Africa and were sounded off on our way by the official hornblower. His job is to blow his long horn at each arrival and departure of aircraft. The horn blowing dates back

*Not the best of pictures agreed, but seen coming out of the Manchester murk in a somewhat unusual configuration is Jimmy Harrison aboard the prototype Andover, G-ARRV, 'down low, one stopped and gear up'*

hundreds of years before man had ever flown and to the time when merchants with their loaded camels came right across the Sahara Desert from the cities in the north such as Algiers and Tangiers, to trade with the people of Kano. As the exact day of their expected arrival would not be known, a designated man would watch the horizon to the north and when he saw the sandstorm caused by the camel train approaching, he would blow his horn to tell the people in the city to get prepared for several days of trading with the merchants from the north.

Back in the UK, after some weeks of analyses of the flight test results and presentation of reports to the Civil Aviation Authority, the Series 2 H.S.748 was granted it's Certificate of Airworthiness and went on to fly for airlines all around the World.

## CHAPTER TWELVE

# The 12.30 to Delhi

During the period in the 1960's when Hindustan Aeronautics Ltd. were building HS 748's under licence, Gordon and I made frequent journeys to and from Kanpur via Delhi. The first stage on the homeward journey was in our regular taxi.

This particular day was very warm and we sweated as this old Austin Ambassador (a four-wheeled version with probably some 500,000 miles on the clock) trundled slowly up to the entrance of a small bungalow. On a first encounter one might be forgiven for thinking the driver had mistakenly brought them to the wrong side of the airfield, to some deserted outbuilding.

Gordon lifted his sweating posterior from the plastic covered back seat.

*"Here we are again, Kanpur International, well done Stirling!".*

Our driver gave a look of relieved satisfaction.

*" You come back soon, Mr. Gordon?".*

*"Oh I would think it quite likely, Stirling".*

We entered the small waiting room and checked in at the counter. The room was bare except for hard chairs around three walls, above which were a few posters extolling the virtues of flying with Indian Airlines.

It was mid-day and it was hot. We looked forward to some respite in the cool cabin of a Fokker Friendship expected in one hour's time.

The well-meaning authorities had in fact provided some air-conditioning in the waiting room. It consisted of a large fan in the back wall to draw in air from outside. On the outside of the wall was a steel frame over which was draped a piece of sacking. We had learnt from bitter experience never to sit under or near the fan. About every fifteen minutes or so an employee, you might say a 'punkah-wallah', would throw a bucket of rather unclean water over the sacking. The result was that anyone sitting under or near the fan received a very unwelcome dirty showerbath. So, despite the heat, we sat down well away from the air-conditioning. We were waiting for the one flight a day to Delhi where we would transfer on to a flight to the UK. Waiting time is the slowest time and it always seems that the hotter the day, the slower goes the clock!

As we waited we studied the other potential passengers and the few comings and goings. Being frequent travellers, this was a pastime we indulged in quite often and with a little imagination and a sense of humour it can often be quite entertaining at times. There were some twenty or so people in the waiting room and only a couple of seats unoccupied, when in came an Indian Air Force officer. We would not have taken any particular interest in him but for the fact that he showed obvious signs of being put out at having to wait in the same room as ordinary Indian people. He was wearing his No.1 uniform, perfectly cleaned and pressed. He stood near the entrance door obviously disdainful of taking one of the two remaining chairs and sitting with the lower classes, as he saw them. He

*748 Series 2 prototype G-ARAY in company colours.*

showed signs of narcissism and kept adjusting his uniform and preening himself. Gordon and I both loved Indian people but we did not care much for the look of him.

After some while, the standing and the heat must have got too much for our military friend and he reluctantly walked across the room and sat down on one of the remaining vacant chairs, right below the air-conditioning fan. Gordon nudged me and we could hardly contain ourselves while we waited for the next dose of air-conditioning. It came, a large spray of dirty water all over the poor chap. He jumped, nearly halfway across the room, his face becoming a study of shocked disbelief. He looked at his uniform now splattered with wet and dirt, he looked back at the fan on the wall, he wanted to kill someone, but whom? He then turned and looked at the two clerks behind the check-in counter.

We could not say whether or not he did kill them and perhaps the poor 'punkah wallah' outside, as at that moment word came that our aircraft was approaching the airfield. Everybody was relieved to leave the scene of the forthcoming massacre and go outside.

We stood outside in the midday sun and fixed our eyes on the speck high up in the distance. It gradually grew nearer and bigger but to our expert eyes, not lower. We continued to watch as it flew over the airfield at about 7000ft. and away to Delhi.

Gordon and I looked at each other, picked up our luggage and went in search of a taxi.

We would try again tomorrow.

# Collective moments in Uttar Pradesh

Uttar Pradesh is a state on the northern frontier of India and is the most densely populated area in the country. Some 110 million people live there added to which there are an estimated 2 million pilgrims passing through annually.

The all important river Ganges runs through the entire length of the state from West to East, finally entering the sea at Calcutta. On its way it passes the renowned historical cities of Lucknow, Kanpur and Benares where the Hindus bathe in its waters to receive blessings in their afterlife.

*The Hawker-Siddeley Test Team at Hindustan Aeronautics Ltd. in Kanpur, with 'Chittaurgarh' a locally-built 748 behind.*

Our work took us on several occasions to Kanpur (called Cawnpore during the days of the Raj). At a factory there the Indians were building our twin-engined 748 aircraft under licence for operation on internal routes by Indian Airlines.

## TALES OF THE CHESHIRE PLANES

The purpose of these visits to the aircraft factory was to assist technically with the flight testing. There was nothing very difficult in these tests, but in trying to achieve a high performance standard when climbing with one engine shut down, little things sometimes matter.

On one particular flight, each time the starboard engine was shut down and the controls re-set for the single-engined climb, the altitude and airspeed instrument readings changed abruptly. This was not normal as the instrument system is designed so that small sideslip effects are cancelled out. After two or three attempts, which all showed the same effect, we aborted the tests and landed back at the factory airfield. As usually happens after a test flight there was a gathering of ground staff to meet the aircrew as they disembarked from the aircraft. The tests and any problems with the aircraft are discussed on the spot and any maintenance work or other changes to the aircraft agreed before further tests are programmed. On this occasion of course, the talk centred entirely on the abnormal behaviour of the instrument system.

The static air pressure supplied to the altimeter and airspeed indicator comes via small holes, one each side of the nose of the aircraft. These holes are known as static vents and I just happened to be looking up at the one on the port side of the aircraft when I saw something that made the whole cause of the problem become clear - or rather unclear. A wasp had just alighted at the hole  and had gone inside. It turned out to be a mud wasp and as the name suggests, they build a nest of mud in any suitable hole or pipe. The mud nest was cleared out of the pipe and the next time we flew all was normal again - though the wasp was probably a bit upset!

If hotels in Europe are not very pleasant to stay in we rate them as 'One Star'. The hotel in which we slept in Kanpur, I am afraid, did not rate any stars. I specifically say 'slept in'  as there were no other facilities and if we had eaten there, I do not think we would be here today to tell these tales.

Fortunately, we had a Company Representative living permanently in Kanpur. He resided in a large very comfortable bungalow in the grounds of the old Cawnpore Club. The setting was one of beautiful gardens complete with grass tennis courts and a swimming pool. Peter, our representative, employed a very accomplished cook who could prepare any meal, Indian or European, even down to treacle pudding and custard. He was a Muslim and although living in a predominantly Hindu area, he was not afraid and in fact was proud to extol to us the virtues of his religion. So we had all our meals at Peter's bungalow and just slept at the hotel.

There were however, spare moments at the hotel, when for want of something better to do, we would sit at our balcony window and watch life in the busy street below. It was a main street and right below our window was a roundabout. A turning opposite us was a road leading down to the river.

When an Indian dies the body is either burnt on a funeral pyre, or if the relatives cannot afford the wood for the fire, the body is put into the holy river Ganges. If one gazed out across the river one could often see two or three kites apparently standing on the water. In fact they would be stood on a body floating just below the surface. The birds would be enjoying a good meal until

*A room with a view... all the traffic in Kanpur seemed to go around this roundabout!*

there was not enough flesh left to maintain buoyancy and the bones would sink.

Members of the H-S test team beside the swimming pool at the Cawnpore Club. Charles Masefield is 3rd from the right, with the author standing!

We were watching from our window one afternoon when just such a poor man's funeral party approached. It consisted only of four men carrying a dooli (a stretcher-like arrangement consisting of two poles with a piece of canvas stretched between). The body on it was covered with just a piece of sheet - no flowers and no mourners. They turned the corner of the road towards the river and it seemed that they considered a rest was needed. Putting the poles down on the road, they sat on the curb for their union-negotiated break.

At the restart of their trudge to the river there was a slight lack of co-ordination. One man dropped his end of the pole and the poor body slid from under the sheet on to the dusty road. It certainly was not a very dignified funeral, and I would not have included this story but that I thought afterwards, even if that man had been given a funeral befitting a rajah or a king he would not have known any different.

As one seldom sees working elephants on the streets of England, the sight of one of them in India was always one of the highlights of our time there. When we did see one, there was almost childish excitement and a rush for cameras. These magnificent beasts generally have a large bell hanging from their necks and so one usually gets some notice of their approach.

As we were looking down from our hotel window on another occasion, an elephant approached with the mahout (its driver) sat astride its neck. It stopped right below our window and the mahout dismounted and disappeared from our view. After about five minutes he reappeared. We could not hear but presumably he said something to the elephant, which lifted its nearside front leg to form a step up for him. Grasping a large ear, the mahout pulled himself up on to the horizontal part of the leg. Just at that moment, a friend passing by stopped to talk to him. We watched entranced, as the two men were in conversation for about ten minutes, the mahout standing on the elephant's upheld leg throughout. The leg held in the air never moved until the mahout was once more back astride its neck. What a wonderful animal !

*It was always fascinating to watch elephants at work...* One morning, as we were having breakfast in Peter's bungalow, the ding-dong of brass bells suggested the approach of several elephants. Bacon and eggs became secondary to grabbing cameras and

rushing out into the garden. Three elephants were coming down the lane adjoining the garden. The mahouts were quite happy to stop to allow us to take some pictures and we in return, were pleased to offer them some small gratuities. For my offering, I offered up a ten rupee note which is very small and when folded must have been no more than half an inch square. At a word from the mahout, the elephant took the tiny folded note in its trunk and passed it up to him. As Gordon remarked afterwards, it is wonderful to think that these huge animals can handle anything from a tiny piece of paper to a 20ft tree trunk.

The Indian army mutiny of 1857 was a dreadful event for both the Indians and the British and there were particularly bloody happenings in Kanpur. Just outside the city there is a sandy road running down to the river and this road is sometimes referred to as Massacre Gap. It was down this road that British Army officers and their families were forced to go, after having been dragged out of the church opposite, by the rebelling Indian soldiers.

The story goes that the families were offered boats in which they could escape across the river. This means of escape was thankfully accepted but when the boats were about thirty yards out from the bank, the Indian

*A commemorative screen to the victims of the 1857 Mutiny at Kanpur.*

*It was in attempting to cross the River Ganges at Massacre Gap in small boats such as this that many British Army families lost their lives in the 1857 Multiny.*

soldiers raised their rifles and opened fire on them - hence the name given to little sandy road down to the river.

When one walks down the road today, one is greeted not by riotous Indian soldiers but by a large flock of white geese, a family of monkeys and some wrestlers. You are offered a boat ride and there is a good chance that you will not be shot, providing you pay of course! There is a small Hindu temple complete with deflowering tools. These implements were used in the past, I was told, on girl virgins, though I am not sure why, as there is a far more pleasant way to do it.

There is also a canopy-covered wrestling sandpit, if

*The Hindu temple at the end of the Massacre Gap.*

one should feel the urge to take some revenge for the Indian Army revolt but then one should reflect that it might be the lithe Indian wrestlers who would care to be revengeful, bearing in mind the British Army's actions in putting down the mutiny.

All is peaceful and friendly now and if you are ever in Kanpur, it is well worth a visit.

Travelling in and out of India, and sometimes to attend meetings, often found Gordon and myself in Delhi. We stayed in the modern hotels in New Delhi but to see more real life and for interest we would often go into old Delhi. I remember one particular street where one of us was always asked by a small boy or girl:

*"Shoes clean Sahib?"*

*" Er-er-er no thanks they are fine"*

*" No Sahib - see!"*

From somewhere a dollop of a white creamy substance was on one of our smart black shoes. So out came the brushes and cloths followed by a couple of coins out of our pocket and a broad smile on the face of the little boy or girl.

We were quite amused by this, for though it happened two or three times to one or other of us, we could never spot who threw the creamy substance on our shoe in the first place.

In the bar one night we devised a plan. The next time we went down that particular street I would walk some twenty paces behind Gordon to spot how the system worked. Well, I failed to spot anything being done to his shoes, but a little girl stopped me and said

*" Clean your shoes sahib?"*.

You can't beat the system!

On another occasion we were in Delhi for a meeting with the Chief of the Indian Air Force. After waiting all morning in our hotel to be informed as to the time of the meeting, we eventually learnt that it would not be that day. So, discarding our formal clothes and putting on cooler ones, we ventured out into the city. After some pleasant sightseeing and a little shopping we ended up on the roof garden of the apartment where the Rolls-Royce Representative (India) was living. It was hot and the most humid afternoon I had ever experienced, the start of the monsoon period being upon us. A very pleasant session ensued and the fact that we were soaking wet just from the humidity was easily countered and forgotten by washing down large 'G and T's'.

No doubt the party would have gone on well into the evening but for one thing - the monsoon broke. One sees heavy rains in Europe but there is nothing that I have seen that quite matches the monsoon rains. By the time we had set off in our car to go back to our hotel the streets were already flooded to a depth of a foot or so. Our driver coped quite well, except that he collided with another car. It was by no means a major accident, but to the two drivers it was a matter of honour and it was quite a unique sight to see the two standing in the water, arguing as to whose fault it was.

When we finally reached our hotel and entered the foyer looking like drowned rats, our Director was there hopping up and down.

A typical street scene in Kanpur, with double-decker buses and English cars.

*"The meetings on again, quick, go and get changed, we've got ten minutes to get there!"*

We dried ourselves as best we could but our dry clothes were wet within five minutes of putting them on. Only a few minutes late, we drove up a sweeping drive to the front door of a very large house. In its day the house must have looked magnificant, but now the gardens and the house looked drab. We were shown into a large drawing room and there we sat in suits and ties, feeling wet and uncomfortable, but so very formal. After about ten minutes of this discomfort the Chief of Staff entered the room - he was dressed in an open cotton shirt, shorts and sandals without socks. He looked so cool! *"Sorry to keep you waiting, gentlemen".*

When one visits a foreign country for the first time to do a job of work, it is made a lot easier if you have a resident agent or representative to meet you and help you with transport, contacts, accommodation etc. Sometimes the resident's wife is also only too willing to guide one on shopping and sightseeing, but there is nothing more annoying and embarrassing than the woman living abroad for the first time in her life and who, after only two or three weeks, thinks she knows it all.

Mrs 'X', I will call her, was just such a woman. Unfortunately we agreed to a trip into town (Kanpur) before we 'wised up'. She insisted that she must take us to the music shop and in the end we gave in. Five of us set off in a taxi - the woman and her 17 year old son, one of our company's directors, Gordon and myself. Winding up a very narrow street, the type with an open drain running down the middle, we stopped at one of the small shops fronting on to the street. It was so cramped inside with the two Indian shop owners and various musical instruments all over the floor, that thinking back, I have never understood how five of us managed to get into the shop!

The whole episode then became more and more embarrassing. None of us had indicated to Mrs 'X' that we had the slightest desire to buy anything but she had the shop owners show us everything. Then she insisted that we be given a recital on one of the many sitars. The local sitar player lived somewhere up the street and had to be sent for, all on the expectancy of a sale. He arrived after a few minutes and played two pieces, very Indian, and as far as melody went, the second piece was indistinguishable from the first. (Travelling back later in the taxi, the 17 year old son of Mrs 'X' announced, possibly to relieve the tense silence: *"I much preferred the second piece he played!"*)

The sitar recital having finished, there was a very awkward silence which was broken suddenly by an awful crunching sound. Our Director had taken a step back, trodden right on top of a musical box and crushed the flimsy wooden construction to pieces!

Our embarrassment was complete - God save the traveller from the Mrs.'X' s of this World!

## CHAPTER FOURTEEN

# When you have to go...

---

### 1 - Cairo

Apart from your health, there are three things we considered vital to the traveller abroad: your passport, money, and access to a toilet when you need one. Pre-travel preparations must take care of the first two but for the third one, depending where you are, you often have to take pot-luck - no pun intended!

In our experience, one feels least assured and comfortable in Arabic countries. There, a lot of the male population seem to reason that as long as there is a sandy patch, who needs a toilet!

Such scenes can be very off-putting, particularly when you are sitting by the open window eating your breakfast! My apartment in Egypt was in the Giza district of Cairo, in sight of the Great Pyramids but more important, only a couple of 'three woods' away from the Mena House golf club.

Although this was quite a pleasant area to live in, it necessitated that I drove 15 miles each way to Helwan where I was working at that time. In due course, wear and tear took its toll on the car's silencer and with the help of an Egyptian friend I sought out the Peugeot agency in Cairo. At the agency I was met with a look that was a mixture of surprise and resignation when I asked for a new silencer.

*" No, no, we do not have this "*

*" It's for my Peugot 504, not an Avro 504..."*

...but this reply was wasted on him. However, to be fair, he was helpful and told me to go to 'such and such street (the name of which escapes me now) where:

*"They  do very good exhaust for all cars".*

I thanked him, and Muhammed and I left.

When we finally got to "such and such" street, which once again, I would never have found without Muhammed's help, it was quite a wide street with shops and various places of business on both sides.

The exhaust business was located, just a small workshop fronting on to the pavement. We found that they did not sell exhaust systems but made them while you waited. I got the usual estimate on how long to finish it.

*"We do it very quick for you Mr."*

But wait we did, some 4 or 5 hours as I recollect, standing about on the dusty pavement while they cut and welded inside and worked on several cars on the pavement and despite Muhammed's pleadings to them, we did not seem to get any priority at all.

After a coffee and one or two 7 Ups, it was inevitable that an old man like me with a slight prostate problem, needed the"little boy's room". Muhammed enquired for me in the workshop and then pointed to the darkest corner of the room.

I had been obliged to use some rough and dirty toilets around the World but never anything quite down to this standard. It would have made a realistic setting for a horror film. The walls and floor were black with greasy filth and the hole in the floor was a utopia for crawling livestock living on decades of deposits. But it was when I lifted my eyes to the ceiling, that the full horror of the place struck me. It was covered with a carpet of dusty cobweb which looked to be at least 2 to 3 inches thick. In one corner sat the owner, the largest and the most frightening looking black spider I had ever seen or want to see again. I looked respectfully at him while he sat there and weighed me up - would his cobweb take my weight?

Never have I finished so quickly!

## 2 - Khartoum

In September 1950, the Shackleton Mk.1 Maritime Reconnaissance aircraft was flown to Khartoum for it's tropical flight trials. It was basically an A.&A.E.E. (Boscombe Down) operation but I was lucky enough to be chosen as the Avro representative to go on what was to be my first overseas trial. We stayed in the RAF Officer's mess in Khartoum and were made very comfortable despite the very hot weather.

One evening, while enjoying a game of snooker, I suddenly received the painful onset of the Pharaoh's curse. I cannot remember now what I had been eating

*The prototype Shackleton M.R.Mk.I at Woodford before departure for Boscombe Down and later Africa.*

*The Shackleton tropical trials team with the aircraft at Khartoum.*

# WHEN YOU HAVE TO GO...

## TROPICAL ATTIRE - 1950's STYLE!

*Left: The author - improperly dressed - with RAF officers from Boscombe Down.*

*Below: Our group (author 2nd from right) half way across the Nile on a hunting trip in the Sudan.*

but for sure there was only one more course to take and that was post haste in the direction of the toilets. These were a row of wooden cubicles situated a short walk away from the Mess. It was 'dry' sanitation of course and there were no lights that I could find in a hurry. It was around midnight, pitch dark and not a sound as I sat, beginning to feel relieved and relaxing into a sort of dreamy trance. All of a sudden, in this eerie silence, there was a bang, followed by a scraping sound and the receptacle under my wooden seat was pulled away backwards. The shock was nearly more effective than the dodgy food I had eaten.

On emerging from the cubicle I saw that the

intruders were two Sudanese gentlemen from the public health department. They were removing the receptacles through flap doors at the back of the cubicles and putting them on to a flat vehicle pulled by a camel. Their approach on the soft sand had been absolutely silent. I had to hope that Pharaoh would send his curses earlier another time!

## 3 - India

During one period in India, we were invited to visit an engine factory in Hyderabad where they were building the Rolls-Royce Dart engine under licence.

While we were being given a conducted tour of the plant, the dreaded 'curse' struck me and it was of such an order that it could not be denied. I had a quiet word with the gentleman showing our party around and he directed me to the nearest place for relief. I say 'nearest', but for someone in my state of distress at that moment, it seemed a marathon away. Finally I reached it and found it to be of quite VIP standard, clean white tiled cubicles with the usual hole in the floor and indented footprints either side for a steady stance. In fact it was so inviting, that some five hundred or so mosquitoes had taken a liking to it and were circling the hole at the very inconvenient height of two feet or thereabouts. It was not a time to be squeamish and he who sends the 'curses' cannot be thwarted!

I lived to tell the tale and a number of mosquitoes died in the action, though I did not bother to count how many!

## CHAPTER FIFTEEN

# The Miracle

If an aircraft type is intended for operation in all parts of the world, including from very high airfields in hot climates, then its performance must first be measured in these extreme conditions.

Before going any further, here is as good a place as any to explain the assorted 'prefix' changes that appear with the 748 design. The original design came from the Avro stable - hence the 'Avro 748'. When Avro became the Avro-Whitworth Division of Hawker Siddeley Aviation back on 1st July 1963 it was re-designated the 'H.S 748'. Later, HSA itself became part of British Aerospace and the airliner was re-named (again) the BAe 748! But to get back to the story...

The venue chosen for these 'hot and high' tests on the twin Dart engined HS748 was La Paz in Bolivia. The airfield there is some 13000ft. above sea-level and is only about 16 degrees of latitude below the Equator. The length of the runway is a comforting 2.5 miles.

We picked a test team of ten to go to La Paz. The two pilots and the two ground engineers travelled in the test aircraft around the North Atlantic via Iceland. The six making up the technical team flew by commercial airlines, planning to arrive at La Paz two days earlier in order to make some advance preparations. Well anyway, that was the idea.

In the event, on arriving in Miami, we found to our secret delight that there were no ongoing flights to La Paz for three days. So we contacted the test aircraft via our base at Woodford and had to wait two days in Miami. It was a tough two days, I can tell you! However, the local people were very kind, offering us drugs, guns, hookers and in fact anything they thought we needed. If only we had brought more money with us!

Our test aircraft had received our message, picked us up and we flew on towards South America. After night-stopping in Panama, we reached Lima in Peru.

Here the welcome was quite riotous! On enquiring of the hotel porter as to which restaurant or night spot could be recommended, he replied that unless we wished to return full of holes, the hotel we were in was it! So that evening we contented ourselves with the facilities in-situ and went to bed to a serenade of gunfire from somewhere outside in the city. Luckily none of us were hit and the next day we got airborne again.

We reached La Paz and as we circled the airport prior to landing, the aerial view was not very reassuring. The runway looked suitably long for our tests, but to the sides and beyond the extremities were strewn the remains of Constellations, Flying Fortresses, DC-4's and other vintage aircraft which, for one reason or another, had not made it. We learnt later that they were nearly all used as cargo aircraft hauling meat carcasses from Argentina - cattle, not human!

Our tests, mainly single-engined take-offs and landings, were completed satisfactorily in about two weeks and the aircraft departed on its way back to the UK. Once again the six of us left behind had to wait a

*Our Vinten f47 Take-off recording camera is up on the roof of the Fire Station at La Paz Airport, Bolivia. Derek Black, Gordon Fox, Martin Garland and the author stand guard over International Fire Truck No.3.*

*The author - looking partially like a Bolivian cowboy - stands ready to use the sighting device used to assess take-off distance to fifty feet above the runway at La Paz.*

couple of days before we could get seats on a flight back to Miami. We usually managed to arrange it this way, so leaving a couple of free days to see some of the sights of the country.

The two young Bolivian drivers who had ferried us each day between our hotel and the airport, proposed a trip through the Andes mountains. All eagerly agreed, and we set off early the next day in two fairly old Jeeps. To enter the mountains one first has to book in and then pass through a police gatehouse and we were told that this entry is closed at 6pm each day and no

*The police gatehouse at the entrance to the Andes Mountains from La Paz.*

further traffic is allowed into the mountains.

At first it was a reasonable road as we drove up to about 16000ft. with snow covered peaks all around us, some of which reach to 23000ft. We photographed this majestic scenery and the llamas grazing on the lower grassy slopes.

*"See train...?"*

our driver said, pointing across the valley to what looked to us to be a sheer mountain slope.

*"Railway, what does he mean?..."*

someone asked. However, after a couple of minutes of intensive staring, it was there, a zigzag line going left, right, left, right up the mountain slope.

After a couple of hours the road, which had now become a narrow rough track, started to descend. We stopped twice more, once to   photograph a spectacular waterfall and once to wait for workmen to finish clearing a landslide.

Finally, when down to about 6000ft. we reached our destination, a village that I believe was called Coroico. It was now pleasantly warm with the sun shining brilliantly on the whitewashed walls of the houses. We ate our picnic, drank some very strong coffee and massaged our backsides, sore from a very rough ride on the very hard bench seats in the Jeeps.

The long drive back was not so pleasant. As the road climbed higher and the sun slipped down behind the mountain peaks, we got colder and the bench seats got harder. I had instructed the drivers of the Jeeps to stay together but this had not happened. Gordon and I were in the second Jeep and the first one was well ahead and out of sight. We had just about reached 16000ft. again when we got a puncture! On our enquiring as to where the spare wheel was stowed, our driver looked blank, not surprisingly as there wasn't one. It was nearly dark, it was cold and we were tired. We were stuck on one of the highest mountain roads in the World, no habitation in sight, no passing traffic - nothing. All our minds went blank as to what we could do and more so when we remembered that no more traffic would be allowed to pass through the police gatehouse at the entrance to the mountains.

On a tarmac road we could have struggled along with a flat tyre but on this rough track it is doubtful whether the tyre would have lasted more than a couple of miles.

We were just despairing of the situation when up the road in the gloom and coming towards us, we perceived the figures of two men. Wherever had they come from? Whatever were they doing strolling along this desolate road? We had not seen a single soul for the past three hours. When they reached us they spoke to our driver in their local tongue. It turned out that they were a gentleman and his manservant out for their customary evening stroll.

*Andean roads - fit for mountain goats only... not the place to get marooned!*

They told us to remove the wheel and follow them. This we did and over the brow of the hill we came upon a large house surrounded by a courtyard. Once inside, the manservant took the wheel and from that moment on we could only stand and watch.

There was virtually no conversation as our driver had only a few words of English and none of us spoke Spanish. However, we did manage to pick up that the manservant's name was Carlos. He stood for some moments looking at the wheel and it's deflated tyre then it seemed that a thought suddenly came to him and he went back out into the courtyard. We followed him out and over to a large dark object in the far corner. It was covered in thick layers of dust and dirt but on a very close inspection it turned out to be a very old Pontiac which must have lain there for twenty years or more. After a struggle Carlos opened the boot and rummaged inside with the aid of a torch. When his head emerged again, his face for the first time showed a hint of his feelings as he held aloft two tyre levers and a hand pump for his master to see. Returning to the house he soon had the tyre and the inner tube off the wheel and had located the hole - it was quite a large one. From an old tin he got from a drawer he brought out what turned out to be his one and only repair patch. Holding it over the hole in the tube he shook his head. We all shook our heads, the patch covered only half of the hole. Carlos looked up at his Master with a sort of 'forgive me' look . It began to look like a pyjamas and toothbrush situation.

After a very thoughtful pause by all present, there was a sudden hurried exchange in Spanish and Carlos and our driver ran outside to the road. We followed and on reaching the road could hardly believe our eyes. Approaching along this dark mountainous road, on which we had  hardly seen any other vehicles all day, was a bus. Carlos stood in the middle of the road

A pause on the road over the Andes so that Gordon Fox can take pictures!

waving his torch and the bus pulled to a stop. He held a short conversation with the driver and then with a wave of thanks the bus trundled on its way.

Once again inside the house Carlos, held up a large repair patch to show us, this time with almost a grin on his face. Again, we bystanders could hardly believe it and as we have often discussed in the years since- what were the chances of buying such an item on that road, at that time of night and at that particular moment. Had the time sequence of our puncture and subsequent events taken place a few minutes later, the last bus would have gone by, or perhaps for other reasons, would never have come by at all. The patch was soon affixed, and the tube and tyre put back on the wheel. Carlos now took the very ancient looking hand pump and started to inflate the tyre. It was just about three quarters inflated when the poor old pump, which must have been still in a state of shock at being dragged out to do some hard work after all those years, collapsed in half. It was enough for the pump and luckily just sufficient for the tyre which went back on the Jeep. It was difficult to thank our hosts and saviours adequately but we did our very best and I think they understood how we felt.

After about an hour's driving we reached the police gatehouse. The first Jeep and its occupants had been to the hotel and had then come back to the gatehouse when we did not arrive. They looked relieved when they saw us for apparently their anxious wait had not been helped with the tales of banditry in these mountains after dark. *"Is that all...?"* was the reply when we told them we had stopped for a puncture, *"...we thought you must have been in real trouble!"*

We were tired, cold and hungry and in no state to try and expand on our excuse; as far as we were concerned we had witnessed a Miracle.

# Customs and Immigration

## Anything to declare?

There are several advantages to flying in the Company's aircraft. Firstly there is just your own team of chaps on board who all know each other and who are all there for the same destination and purpose. You can choose and plan your own times, routes, stopovers and hotels to fit an overall schedule. An additional perk is that you can bring your own choice of food for on-board meals and in our team we had a very good navigator who took it upon himself to do all the planning including the provision of food and drink. As a result we always travelled privately, comfortably and gastronomically in style.

It had been one of these comfortable trips when we landed at Delhi one day. Being a private aircraft, air traffic control directed us, not to the normal passenger terminal area, but to a parking spot on the far side of the airfield. This was no problem, as a coach arrived and took us over to the main terminal building where we passed through passenger immigration and customs and then were soon on our way to an hotel for the night.

Now, although we bodies and our personal luggage had cleared customs into India, our aircraft with its contents had not been cleared. Of course, we had nothing to declare with respect to its contents, but a foreign country has to imagine that it carries illegal arms, bullion, drugs and the like and so they want to inspect it. So early the next morning, Charles Masefield as captain of the aircraft and myself arrived at the airport customs office to ensure that they lost no time in sending an officer out to the aircraft to inspect and clear it for onward flight.

Well, the senior customs officer said that all his men were busy at that moment. We spent several hours in that hot little office, asking politely, pleading and at times threatening as best we could, but all to no avail. Finally they admitted that it would not happen that day and would we please come back tomorrow. This we did, and again on the day after that until finally, on the third day, an officer was at last detailed for the job.

Transport was arranged and we proceeded out to the aircraft. Now it had stood, all doors closed, for three days in the sun, in a temperature of around 100 degrees F. Opening the door was like opening an oven door to pull out the Sunday roast. Unfortunately, or perhaps fortunate for us, things left in the aircraft included a substantial amount of food - ham, chicken, cheese, milk and other nice perishables. In addition, there were about thirty small packs of butter which were originally on an overhead luggage rack but were now a yellow stream down the inside trim.

The customs officer strode up the steps and went to go inside but after one breath, gasped, turned and came down the steps quicker than he went up. *"Cleared"*, he said, signed the paperwork and was off.

**Where were you on the night of...?**
As a holder of a British Passport, one seldom has any difficulty passing through immigration at foreign airports, particularly Commonwealth countries. So it comes as a bit of a shock if the immigration officer does not follow the usual routine of a look at your face, a look at the photo, a nod and hands the passport back to you.

Once, when entering Delhi airport for the umpteenth time, Mike, one of our team, did have such a shock. The immigration officer took his passport and on this occassion studied it with more than the usual casual interest. After what seemed an age, he turned and spoke to another officer standing behind him. Turning back to Mike, he said: *"Please would you go with this officer to the office, there is a query on your passport"*.

Mike, a bit surprised, followed the officer into the office where behind a desk sat a more senior looking man and another stood beside him. There was a discussion in Hindi between them and the senior officer studied the passport now in his hands. Mike, getting a little impatient, asked:

*"What is the query on my passport ?".*

The senior officer spoke:

*" You were here in India this time last year?".*

*" Yes ".*

*" Why did you kill your brother?".*

*" I haven't got a brother".*

*" Exactly, because you killed him".*

Of course, Mike was somewhat taken aback. However, just at that moment, I and two other members of our team entered the office to see what was going on. The matter was soon cleared up and we, together with a quite relieved Mike, left the office.

So, if someone with the same name as yourself is murdered while you are in the country, or are entering the country, have your alibi ready!

**That your case, Sir?**
Some days at airports are more eventful than other days and it is usually when entering rather than when leaving a country that any problems arise. The day I left Cairo after living there for nearly four years, was the exception. Nothing very dramatic for me, as it turned out, but a day when it could have been. As it was, it did not turn out to be a good day for a certain other gentleman.

My wife had returned to England ahead of me and I was left with the job of packing for the final return.

Luckily I was fairly practised at it, but leaving an apartment after nearly four years stretched me a bit. On the final morning I entered a taxi with seven large suitcases, golf clubs and hand luggage. With me was a caddy from the Mena House golf club who had agreed to help me into the airport for a handful of piastres.

One item in the packing had caused me some concern. This was a solid brass mortar and pestel that my wife had bought in the souk - I did not dare leave it behind! Together they must have weighed some 30lbs. I wrapped them in bath towels and put them in the middle of one of the suitcases and I was concerned because that case felt very different from the others when it was lifted.

After struggling into the airport with the help of my caddy friend, my concern was justified when one of the airport baggage handling staff called the attention of his supervisor to that particular case. I was asked to open it but when he saw the cause of the unusual weight he grinned and said: *"Well you did not look like a hijacker but we do have to be careful"*.

It had been a struggle and I was relieved to settle down in my seat, though little did I know how close I was to having to get off the aircraft and pick up all that luggage again. After some five hours snoozing above the clouds we were looking down once more onto green fields and descending into that always welcome Manchester drizzle.

With so much luggage I did not have the face to go through the green channel, and pushing one trolley and pulling a second one, I headed through the red. I knew the man on duty from numerous past occasions.

*"Hallo,"* he said *"Back again? But how did you get out?"*

*" The usual way, by DC10, why do you ask?"*

*" Oh! You obviously have not heard, President Sadat was assassinated this morning and the airport was immediately closed"*.

CHAPTER SEVENTEEN

# Chariots of Africa

(by Gordon Fox)

In 1967 Hawker Siddeley produced a new model of the H.S. 748, the Series 2A. This model incorporated the higher powered Rolls-Royce Dart Mk.7 engines which gave the aircraft a significant improvement in performance. Having measured the take-off distance at Woodford (sea-level), it was then necessary to repeat the tests at an airfield that was considerably above sea-level. The venue selected for these higher altitude trials was Asmara in Ethiopia. The engineers and pilots chosen to form the trials team were looking forward to a couple of weeks in the warm sunshine of Northern Africa, but when they arrived in Asmara they found that morning temperatures were close to 0 deg.C and in fact some of the first take-offs of the day were carried out in temperatures below zero! Even with the airfield being at an altitude of 7,500 ft. above sea-level, these ambients were unusually cold for Ethiopia.

After each day's work was completed, however, the evening's pastimes did warm up quite a bit for the more

The H.S. 748 Series 2A test team outside the Hotel Asmara in Asmara, Ethiopia.

adventurous members of the team. Glancing blows were dealt at the local bars and when thirsts were quenched and sometimes more than quenched, those still not quite warmed up adjourned to the night club.

All these places of refreshment and entertainment were situated in the main street, the night club itself being about half a mile further up the street from the bars. There was no need to walk, as there was always transport waiting outside the bars ready to convey the tired and the incapable back to their beds, or wherever. This transport consisted of a car axle and wheels, on which was fixed a bench type car seat and a four legged one horse-power prime mover between the shafts.

This particular evening one of the even more adventurous of the party, on emerging from the last bar, proposed a chariot race up to the night club, the loser to buy the champagne. Four of the "gharris" described were hired and lined up across the street, each holding two passengers and the driver. By this time quite a crowd of onlookers (we can't say wellwishers) had gathered, and from the shouts of encouragement it seemed that a large grey horse called

*Main street Asmara, scene of one horse-powered drag racing!*

*'Winston' in the pits, recovering from the previous evening's racing.*

Winston was the favourite, and was expected to win. As the starter's flag was raised in readiness for the 'off' the passengers in Winstone's gharri had a smug look of expected success on their faces. The shouts from the onlookers rose to a crescendo, the starter's flag came down and with the driver's whips flailing encouragement, they were off! All except Winston! He refused to budge, despite corporal administering by the driver. By this time the other three gharris were some thirty or forty yards up the street and going flat out. Some desperate initiative was called for! The passenger next to the driver grabbed the whip and inserted the stock end into Winston's rear orifice. Winston moved all right - he leapt about three feet into the air with all legs flailing in a running action! When he returned to earth sparks flew from his shoes reminiscent of a 'Brocks Benefit Night', but it was all too late and he never caught the others up. Anyway, the champagne flowed and a good time was had by all. Since then, many a pint has been drunk as the race has been 're-run' and Gordon always finishes the tale with "guess who won?".

## CHAPTER EIGHTEEN

# Stripping in South Yemen

In addition to the successful twin Dart civil aircraft (748), the Company produced a military version for the RAF, which was called the Andover. In appearance it differed from the civil version in that it had large tail doors and a ramp, which when closed, gave a swept up tail end. In addition, it was fitted with R-R Dart 12 engines, the most powerful in the Dart family, and larger diameter propellers. Inside of course, it was fitted out to RAF requirements for carrying troops, including paratroops, vehicles and military stores. Finished in a desert brown paint, it looked military and right for the job.

Like the civil version it needed to be capable of operating anywhere in the World, but also from airfields and landing strips where the civil aircraft would not normally be asked to go. Some of the strips that the Andover might have to cope with would not be constructed of concrete and tarmac, but the surface would be of stones, sand, gravel and suchlike.

Before placing an order for 31 Andovers, the RAF required a demonstration of take-offs and landings from selected non-paved strips. First they chose

*Company demonstrator and test machine G-ARAY prepares to tackle the specially prepared mud strip laid out at Martlesham Heath - absolutely no problem! Its excellent performance here in February 1962 secured the RAF order for the 748 MF version (later called the 'Andover').*

*Another view of the mud runway at Martlesham Heath that demonstrates the ruts G-ARAY (seen in the background) had to deal with. Its success here gained the RAF order in the face of competition from the Handley Page Herald.*

Martlesham Heath where they prepared a strip of deep soft mud. The 748 did so well off this surface that the order was won more or less there and then.

The Andover, being of a low wing design, left the RAF with a slight question as to whether or not there would be unacceptable damage from stones when taking-off and landing on these rough strips, particularly on the underside and the propeller blades.

*The joint H.S. A&AEE team (author on the extreme left) pose in front of the second production Andover C. Mk.1 (XS595) before departure to South Yemen.*

## STRIPPING IN THE SOUTH YEMEN

The only way to settle the question of possible stone damage was to go and operate the aircraft from some typical rough strips.

The RAF selected the strips to be flown from five places up country in the Aden Protectorate, now of course, South Yemen. We flew our Andover out to the RAF station at Aden and this was our base for the trials. The plan was to fly up country to one of the five selected strips each day and return to Aden for the night. A senior officer warned us that there was bandit/terrorist action going on and that we must be sure to return to Aden before dark each day. If we broke down and could not get back, the RAF would not be able to assist with any sort of rescue or protection. This gave us some food for thought in the planning of our daily programmes. We had every confidence that there would be no problem; but on one day there nearly was.

*Check-In Counter on one of the sand and stone strips.*

*XS595 clears the sandy strip at Neguib in March 1966.*

On each of the five days, we flew out to a different strip. The surfaces on four of them were very stony indeed, and the fifth was sand-covered. Several of us would stay on the ground when we reached the strip to observe the take-offs and landings and to take film records. The trials went very well, but each evening back at Aden, some of us had the back-aching job of noting and marking all the stone marks on the underside of the aircraft that had been suffered that

*Above: A group of locals mix with our escort at the Dhala Strip in the South Yemen.*

*Below: Jim Harrison and the author discuss a problem, out on one of the rough strips.*

day. This was necessary in order that each days assessment of stone damage could be seperated.

On the strips, we hardly saw any of the local population, though on one occasion I had to act like a school crossing lady in order to stop one of the locals and his camel from crossing the strip just as the aircraft was landing.

On another day, we were approached by a party of five locals, headed by, I suppose, an Emir dressed in colourful flowing robes. He held in his hand a cloth bag, which through sign language he offered to us, if we would fly his son back to Aden. It was very tempting, for who knows what the bag contained. However ,

*The Andover contemplates the stones at Wadi Ain. The characteristic 'nose up' pose when the rear loading ramp was down is clearly visible here - so is the size of the stones!*

in view of the tense political situation and the fact that we were flying a military aircraft back to an RAF base, we had to decline the offer.

At the last strip we visited, we found it to be particularly stony - and plenty of sharp ones at that. I believe the RAF classed it as the roughest they had encountered anywhere in the World. Despite this we completed the six take-offs and landings and were just congratulating ourselves on successfully completing the trials, when our engineer dropped a bit of a bombshell.

*" Look at this...",* he said, leading us over to the nosewheel leg. *"Isn't that a beauty?"*

We looked down and there on one of the twin nosewheel tyres was a 2 inch balloon sticking through a bad cut in the outer tyre. We carried spare wheels and jacks.

*"How long to change it?"* asked our pilot.

*Below: Taking a pause from the trials at Neguib.*

*"About twenty minutes if all goes well".*

All didn't go well. The spare nosewheel had gone flat

*Measuring the size of the task! Rough strip trials at Wadi Ain, South Yemen. The largest 'stone' is about four inches in diameter!*

and we did not have a pump to inflate it.

A lot of head scratching now went on, and at the same time we were reminding ourselves of the RAF's warning that they could not mount a rescue flight after darkness fell. After a few minutes our engineer came up with a suggestion...

*"I could cover it over with some strips of strong insulating tape".*

We had a very young Government inspector (AID) with us.

*"You can't do that, it would never survive the take-off on this strip".*

*"Well would you rather stay here for the night and be shot at or captured? What do you suggest we do then?"*

Tape it we did, six lengths one way and six lengths across. We took-off, fingers crossed, and in due course landed at Aden. We all went round to look at the tyre. Not only was it still inflated but the tape strips were still in place, though perhaps just a mite frayed at the edges!

# Sand, Pyramids and MiG 21s

When you ask your boss for a change of job, you sometimes get a surprise!

My 'phone rang. I picked it up - it was my boss.

*"How would you like to go and work in Egypt?"*

*" Well, I wouldn't mind, what am I going to do there?"*

*" How about overhauling MiG 21's?"*

*" MiG 21's! I have never even seen one! I don't know the first thing about them".*

*" Neither do any of the rest of the team out there, you'll soon sort them out".*

*" OK then, when do I go?"*

Apparently, our Company (British Aerospace had now taken over Hawker Siddeley Aviation), were making strong efforts to sell our Hawk trainer aircraft to the Egyptian Air Force. Part of our sales 'co-operation' was, at the E.A.F's request, to send a team of engineers to Egypt to assist them to carry out the 600 hour overhaul on their MiG 21 fighters. While the Russians had been there, little or no expertise had been passed on to the EAF engineers, the manuals left behind were all in Russian and most of the MiGs were grounded for want of an overhaul.

I was met at Cairo airport by two chaps from the team already there, in the early hours of Friday morning, March 1978. They had been up to the airport once before, a drive right across Cairo from Giza to Heliopolis, only to find that my flight would arrive at

The BAe team at Helwan, Egypt.

least seven hours late. They drove me to the hotel where all the team, about 12 in number, were living. It was the Swiss-run *Jolie Ville* situated quite close to the Giza Pyramids. Accomodation there is in twin bedded cabins set in gardens which were once a royal fruit orchard. A very nice hotel and well worth staying at, if you find yourself in Cairo without a hotel already booked.

### 'And the traffic today is...'

Friday, being the day off, I was able to meet the rest of the team and be briefed and 'wised up' to all the things they thought I should know. Travel to and from the place of work was by car and there were two routes, one by the main road alongside the Nile, and the other was by a minor road which avoided having to drive through the city traffic. The minor road was generally used, but I was given some serious advice about using it, which was that, if by any chance I was unlucky enough to knock down one of the villagers, I was on no account to stop. Apparently, a doctor and his wife knocked down a child and, being a doctor, he immediately stopped and got out of the car to help. The child died and so did the doctor and his wife - they were hacked to death! One of the team drove me for the first week until I got a car of my own, but when I did, I can tell you, I have never driven so carefully in my life.

For the first five or six miles the road ran alongside the edge of a canal, with farmland and the occasional

*A view across the canal from the road I travelled to work on - Cairo to Helwan. Never was so much care taken by a car driver!*

village on either side. The little self-built dwellings fronted right on to the road,this on a very narrow road with potholes, some of which were four feet across and a foot deep. Villagers, goats, chickens and other livestock all wandered about on the road with no thought for a terrified white-faced driver, trying to steer his Peugeot in between them and not end up in the canal.

Despite all the hazards, one had a wonderful insight into the lifestyle and work of the Egyptian people living along this road. For instance, two days a week were 'meat' days, and as I drove by I would see them carrying out the butchering on the bank of the canal. At date harvest time, I watched them climb the palms to collect the fruit and build a stockade from branches to protect it as it lay spread out on the ground to ripen. As we went through the seasons, every day there were scenes of Egyptian rural life to be seen and enjoyed. The road was really like a long narrow tapestry running alongside the canal. Without rain (I experienced only two days of rain in the three and a half years I was there), the Egyptians have to live and farm down narrow strips of land on either side of the river or the irrigating canals. Now, some fourteen years later, I was told but find it hard to believe that you can now do the same journey by subway!

**'A bargain is what you think of it...'**

*BAe Engineers helping Egyptian Air Force staff in the MiG 21 hangar.*

Nearing the end of the drive home, there was a tourist type gift shop. They sold all the usual wares attractive to visitors: Bedouin rugs, paintings on papyrus, copper and brass ware, leather wallets and handbags and all sorts of jewellery. I had called in on a number of occasions and had taken visiting friends there to buy their homegoing presents, and so the family that ran the shop knew me quite well and treated me as a friend. Jean's birthday was in a few days time and I called into the shop to look again at a very colourful brooch that I had seen on an earlier visit. I was never knowledgeable with regard to stones, semi-precious or otherwise, but the brooch looked good and I remembered that Jean had liked it when she had seen it. It was not on display, but kept in a drawer.

*" Now Ahmed, let me have another look at that brooch you showed me the other day".*

*"Ah! Mr. John. How nice to see you again. Welcome!*

*Jean negotiating a better price in the Cairo Souk.*

*Welcome! Of course, I have the brooch still because I see that you and Madame Jean like this one. I do not sell it but save it specially for you".*

*" Yes, we do like it and it is Madame Jean's birthday next week. How much do you want for it?"*

*" For you Mr. John and Madame Jean, I do a very special price, £175 Egyptian".*

I put on my dis-interested look. *"No, I don't think so".*

Ahmed put on his shocked look. *"This is a most beautiful brooch, I save it just for you especially".*

Then we went through the usual routine.

*"£125 Ahmed?"*

*" Oh, Mr. John. How can I, but just for you I give it for £165".*

We finally agreed on £150, and off I went with the brooch nicely gift-wrapped.

On her birthday, Jean was pleasantly surprised and pleased with her present. A few days after, she was in the famous Khan el Khalili bazaar in Cairo and taking tea with one of the well-to-do and highly educated shop owners. They knew each other quite well and both enjoyed taking tea outside his shop and discussing life and the world as they each saw it. Jean showed her friend her birthday present and asked him what he thought it was worth. He did not answer right away, but called over a little boy and said something to him in Arabic. The boy went off down the busy alleyway lined with shops. In about 5 minutes he was back. He held out an identical brooch for Jean to compare, and her shopkeeper friend said: *"He has just bought this one for 25 piastres".* I was cross about it. Not so much for being so stupid and being ripped off, but because the deal had been done with someone I thought was a friend. I went back to see my 'friend' and shamed him into giving me my money back. *" Ah! well,"* I said to Jean, *" I won't be caught again...".*

But I was.

*A day off, and some of the BAe team relax with a round of golf at the Mena House Hotel course, Giza.*

### 'Walk your pump Sir?...'

One Friday (the Egyptian week-end and our day off) I drove Jean and our friend Muhammed up to Alexandria, a drive of about 140 miles eachway. Having spent a pleasant day there we set off on the return journey. About halfway along there is a restaurant, where one is glad of a break from the car, and for a couple of cold drinks. We parked the car in the spacious car park in front of the building, and went in for about half an hour.

When we set off again, we had only gone some 400 yards when the engine cut out, and just would not start again. My car, a Peugeot 504, had been so reliable that the fact that the engine had just died so suddenly and completely, led me to think that possibly the fuel pump had packed up. We had just lifted the bonnet and were peering inside, when two men caught us up and asked us what the trouble was. Apparently they were car mechanics and worked at a garage about a mile down the road. They agreed that it was probably the fuel pump and said that they could get one in part exchange at their garage. They removed it and set off to walk to the garage which I could just see in the distance. After 40 minutes or so, as good as their word, they were back and fitted a new pump. We were quite pleased to pay them the £50 they charged us and we were on our way. At first we could not believe our luck, but I could not help thinking that it all seemed a bit too lucky to be true. How fortuitous that the two men just happened along just as we broke down, and that they happened to work at a garage just down the road!

I learnt later that we had fallen for one of the oldest 'con' tricks in the book, or rather on the Alexandria Road. While people are in the restaurant, they disconnect the fuel pipe at the back of the car, and of course, the fuel in the carburettor doesn't take you very far down the road. You are so relieved to receive such fortuitous assistance, that you gladly but unknowingly pay £50 to have your own fuel pump removed, walked down the road, brought back and refitted!

## CHAPTER TWENTY

# Something special

It was the best part of the day. The sun below the horizon and the temperature down to a civilised level. After a day's labours in the heat of the desert, to be sat in the air-conditioned restaurant of a Swiss-managed hotel eating and drinking with friends, was absolute bliss.

There were three of us. David was one of our British Aerospace team overhauling MiG 21 fighters for the Egyptian Air Force, while the other friend was Diana, a freelance secretary working her way up the ladder of occupational pickings in the Middle East. Diana was shortly going to leave Cairo for the Emirates, and three times her present salary.

We were eating in the *Jollie Ville* hotel not far away from the Giza pyramids. It was also the hotel in which we chose to live, being at that time situated beyond the worst of the traffic, apartments, shops, crowds and noise. The food was enjoyable, the conversation happy and the Stella was slipping down with no effort at all. We had just finished the main course and were weighing up the sweet options when David suddenly cried out and his hands flew to clutch his abdomen.

We were startled to say the least, it was so sudden.

*"What ever is the matter?"*

A silly question in retrospect.

Of course he did not know, except that the pain was acute. We thought that perhaps it was just a spasm and would pass. The Pharoahs struck with their 'curse' every so often in Egypt, but never that suddenly, and in any case we had been eating well prepared hot food.

But the pain did not go away and David said it was, if anything, getting worse. We slowly, and for David

*My chalet home in the grounds of the Jolie Ville Hotel, Giza, Egypt.*

painfully, helped him to his chalet and got reception to phone for a doctor who, thankfully, arrived quite promptly. His examination left him in no doubt at all - an acute appendicitis.

*"He needs an operation, "* concluded the doctor.

*"I had better get him booked on the first flight back to the UK"* I suggested.

*"No! No!,"* cried the doctor. *"An operation now, this evening, I will ring the Anglo-American hospital in Gehzira at once. You fetch your car and be taking him down there".*

An ambulance was not mentioned, and so we carefully helped David onto the back seat of my car, where he lay groaning with his head on Diana's lap. I drove to Gehzira, compromising as I thought best between speed and avoiding a rough ride, as every bump was agony for poor old David.

When we arrived at the hospital, two attendants, in white coats that had definitely seen better days,were awaiting us. David was carefully lifted on to a patient's trolley and wheeled into the hallway, where the attendants undid his dressing gown and started to feel his abdomen.

*"Hold on, I think it would be better to wait for the doctor to examine him"* I said quickly.

*"We are the doctors"*, was their rather hurt reply.

*"I do apologise"*, was all I could think of saying.

Diana and I waited until David was wheeled into the operating theatre and then, being now redundant as to helping David anymore, we left.

The next evening when we visited, David was sat up and back with the living again. He felt fine, so after keeping him company for as long as the nurse thought appropriate, we left. We were just descending the stairs when another white coated attendant, or nurse (and here I had to be careful) beckoned to us.

*" Would you like to see something special?"*, he asked.

Not wanting to offend and being slightly inquisitive, we followed him into a nearby room. Each wall was fitted with shelves on which were numerous jars, each containing a piece of flesh in, presumably, formaldehyde. The man lifted one jar down and held it up triumphantly.

*"Mr. David"*, he said, beaming all over his face.

*"Yes, I thought I recognised him"*, I said.

A few weeks later we reluctantly agreed that we had better go and 'do' the Cairo museum. Unless one is a student of Egyptian ancient history, this museum, wonderful in its way, soon becomes a little overwhelming for the ordinary tourist. I'm afraid we were all in this latter category, as we trailed along behind a group and hardly within earshot of the guide. He was pouring out a continuous stream of the names of kings, queens, pharoahs and dates thousands of years BC until one's mind was completely swamped, and rebelled against any further input.

## SOMETHING SPECIAL

The museum is housed in a large four storey building - we never did get to the top. On every floor it is mainly a mass of stone, from the smallest fragment to 20ft. statues. There are other exhibits, of course, but stone predominates and when you think about it, only hard material could have survived for all those thousands of years. We had hung on to our group and tried to take it all in, until someone made a fateful suggestion - what about a cold beer at the Hilton? We were just about to descend the stairs to sneak away, when from a corridor to our right, a policeman in a rather grubby khaki uniform was trying to attract our attention with some Pst! Pst! calls, and beckonings. What can he want, we wondered - is it illegal to leave the conducted party before the finish of the museum tour? Once again, partly due to his insistence and partly due to our own inquisitiveness, we trooped down the corridor to see what he wanted.

" *You want to see something special ?*" he asked in a rather guarded whisper.

" *Oh! not again - all right, what is it?*"

" *Please to follow me*".

We did and entered just another room full of glass fronted cupboards each packed with exhibits. Still beckoning us to follow, he went to one cupboard and took out the 'something special'. It was another piece of stone.

" *What is it?*", we asked.

" *Ah! Nobody knows. It was found in the tomb of King...*" (we never did catch the name ).

We payed up some baksheesh and escaped across the road to the Hilton.

## CHAPTER TWENTY-ONE

# A Philippine incident
## (Gordon Fox investigates and reports)

Statistics tell us that airline flying is the safest form of travel, but from time to time accidents do happen. When an incident is reported, the authorities - that is the airline, the aircraft and engine manufacturers and the appropriate air safety board - must respond and act immediately. First and foremost, action is taken by the emergency services, but as soon as the passengers and crew have been taken care of, representatives of the various authorities must be at the accident site. It is vital for them to inspect and investigate the aircraft before it and the 'accident scene' have been changed or interfered with for any reason.

In 1980 an incident to an H.S.748 of Philippine Airlines was reported and Rolls-Royce immediately instructed me (Gordon) to proceed to the crash site to investigate. Approximately 96 hours later, I was in Manila along with other investigators, attending a briefing on what had happened. According to the captain of the aircraft, he aborted a take-off because he heard a "cough", as he put it, from the port engine and in attempting to bring the aircraft to a stop, the wheel brakes had not responded and the aircraft was only brought to a halt when it ran into the rear of a steel "dumper truck" parked in a yard at the end of the runway. The First Officer had died in the accident and several of the passengers had been injured.

After an early take-off the next day in a King Air aircraft, we had arrived at the crash site by mid-morning and the standard routine of investigation commenced. The site was at Jolo Sulu, on one of the most southernly of the Philippine Islands, where we worked away in a temperature of 30 degrees C plus.

Due to the very early take-off that morning we had missed breakfast, and so lunchtime came and with it

*An aerial picture of the crash scene at Jolo Sulu airfield in the Philippines. The remains of the 748 can be seen bottom centre.*

hunger and a raging thirst. We were escorted to the "Ritz" by six armed guards who informed us that the area was a hotbed of Moslem terrorists. Sitting in the dining room surrounded by the guards was quite an experience - but the arrival of the meal proved to be an even greater experience. We have been confronted with some 'meals' over the years travelling the World but this one had to take the prize. It was the most foul looking chicken stew - not even curried. The local public relations fly had done his job well in spreading the news, and no sooner had the plates been set on the table than hoards of flies descended on us. They came from near and far and I was sure I recognised one or two from Manchester! They need not have rushed because there was no way that we could eat the stuff and expect to survive the day. We made as polite a refusal as we could and settled for a liquid lunch of 'San Miguel' . Our feelings, as we made our way back to the airport to continue the investigation, were not only about how hungry we were, but how soon we could finish our work and climb aboard the King Air that

Two further views of the wreckage of the Philippine Airlines 748 crash off the end of the runway at Jolo Sulu.

The 748 crashed into a dumper-truck in a council yard on take-off.

brought us from Manila, and get back to relative civilisation. But life in Jolo Sulu was not that simple for, to our astonishment, we found that the pilot of the King Air was already airborne on his way back to Manila, and we were stranded there! Before starting on that problem however, I authorised and arranged for

*Philippine Airlines YS-11 fitted with a forward freight door on the apron at Jolo-Sulu with ground handling equipment ready to move the pair of Rolls-Royce Darts back for investigation.*

the "coughing" engine to be removed and transported back to the overhaul base in Manila, and a YS-11 arrived to do this.

Now how to get ourselves back to Manila? On enquiring, the answer we got back was: "NO PROBLEM!" That really worried us, as these are the two most misapplied words in this part of the World. However, the Air Force arranged to fly us to Zamboanga where they told us we could catch a scheduled flight to Manila.

Things were beginning to look a lot brighter, that was until we saw the 'air transport' the Air Force had dug up from somewhere. It was a very small scout aircraft of a type neither of us had ever seen before, and we were a bit too taken aback to bother enquiring. Two

*Trying to take a photograph down the cabin caused the photographer no end of focus problems, but the pair of Darts from the crashed 748 can be seen here.*

of us and the pilot just managed to squeeze in, huddled together with brief cases on our knees. Would it fly? The moment of truth dawned as the pilot applied full power and released the brakes. The aircraft started to move all right but in a circle, spinning round like a top and coming to rest on the grass.

Pilot, switching engine off: " *No problem!*".

Mechanic arrives and fiddles around with the offending wheel: *"No problem!"*.

Pilot climbs back in and gives us a detailed report on the situation: *"No problem!"*.

All this had been carried out in good humour by everyone concerned, except us, sat there wondering what was going to befall us next or even if there was going to be any 'next'.

Throttle fully open again and this time we managed to get airborne. Although it was only a short crossing over the sea, to us it seemed an eternity, with great relief when Zamboanga came into sight. Our scheduled flight departed on time, and we were back in Manila by midnight - to a long awaited meal and several bottles of San Miguel.

The outcome of the investigation resulted in the pilot being 'asked to resign'. Rig and engine testing of all the relevant units removed from the suspect engine showed that they all performed satisfactorily. The cockpit voice recorder demonstrated that there was no abnormality in the sound of the engines during the aborted take-off run. It did however provide undisputable evidence that there had been a total disregard both during the landing prior to the accident and during the fatal take-off run. Most of the voice recording was that of the captain discussing with the stewardess, the spicy merits of a book entitled 'Black Emanuel'. You might think it was a pornographic conclusion to that pilot's career!

# On Your Bikes!

It was 1985, and Gordon and I were already in the last decade we would be employed by our respective Companies (though how they would ever manage without us we could not imagine). Gordon was still upholding the Rolls-Royce tradition at Woodford while I was now based at the Kingston-upon-Thames factory. Previously this was the Hawker factory and I had a shock a few years later when I drove past it. It had, almost unbelievably, been razed to the ground and the site cleared until it was as flat as when it was a green field. Standing at the gate through which one had passed so many times and gazing on the desolate scene gave me an eerie feeling of disbelief - this could not happen - but it had. I had only worked there for a few years but what a heart wrenching blow for those who had worked there all their lives and had been part of the designing and building of such famous Hawker aircraft as the Fury, Hurricane, Hunter, Harrier, Hawk and many, many others.

At the Kingston factory, after my return from Egypt, I was involved in a sales project endeavouring to sell Hawks to the Algerian Air Force. Part of this deal was that we would help to modernise their MiG 21 fighters, hence my participation in the sales team. This involved several visits to Algiers and one to Oran. A rather strange encounter was experienced one day in Algiers when I was sightseeing-cum-window shopping with another member of our team. Having become a little leg weary and reached that 'coffee time', we went into a bistro. It certainly looked the right place for coffee, for on the bar was one of those wonderful large machines that hiss and dispense instant boiling water in clouds of steam. It was quite a large place but was devoid of customers except for five men sitting round a table near the bar and the barman himself. In our best schoolboy French we asked the barman for two coffees, the response to which was a surly, suspicious look, a quick hissing by the machine and the production of two piddling little demi-tasses. Never mind, we took them and sat down some distance away from the five locals who were talking earnestly. The weight off our feet was welcome but the demi-tasses did very little for our thirst and so it was not very long before we approached the not so very happy looking barman for two more. One of the five locals said something in Arabic to the barman which, despite my four years in Egypt, I did not pick up. The response to our request we did understand, it was *'non'*!

Enquiries, in schoolboy French, as to 'Why not?' did not seem to be a prudent next move, and so we slunk off out into the street with puzzled looks. Maybe our accent sounded too Parisian? Whatever the reason, it was certainly not a shortage of coffee or hot water or that the place was crowded out. We had our idea as to what might have been going on at the local men's table and a few years later the Algerian Fundamentalists started their terrorist activities. It is surprising the situations one can innocently drop into. Once in Cyprus, when we asked a taxi driver to take us to a night-club, we ended up in the EOKA headquarters. Fortunately, that was some time after their problems with the British and luckily most of our party looked rather more Greek than Turkish, although we were eyed up with some suspicion.

After some months and several sales presentations I got the feeling that this particular project was not going to get anywhere. I did not have enough to do and became more and more frustrated to the point where I went to talk to the MD, whom I had known in my Woodford days. The upshot of our talk was that I took up his offer of early retirement (I was then 63). A few months after this Gordon also took early retirement and so we were both on the unemployment statistics game.

But not for long! We had often discussed the idea of going into partnership as consultants, and that is what we did. Before we had business cards printed we had to

*The author (right) and Gordon Fox of Rolls-Royce helping Zambian Airways.*

think of a name for the business. After several suggestions which did not quite describe our enterprise in a word, Gordon came up with the name 'Dartech' and it was perfect, for we were going to specialise in the technical and performance aspects of aircraft powered by R-R Dart engines.

We spread the word around the many friends and contacts we had in the aviation world and soon became aware that one piece of advice we had been given was very sound and eventually was borne out in our case. A friend, who was already in the consultancy business, told us: *'It is not WHAT you know but WHO you know and when your contemporaries fade away so will your business'*. So we had to get 'on our bikes' (as Lord Tebbit would have us), and get cracking!

Funnily enough, one of the most absorbing aspects of the business was thinking up ways to obtain jobs and the approach to possible interested parties, mostly airlines. Our friend was right, we got nothing through letters and adverts, all our assignments came by personal recommendations. One thing to do was to visit trade shows, in our case the Society of British Aircraft Constructors' display at Farnborough, and its equivalent in Paris, though first and foremost, to acquaint all our business friends and associates with the fact that we were now available as consultants, and which aspects of the aviation business we were covering. Another factor bearing on the time span in front of us, was that aircraft powered by R-R Dart engines had been flying since 1948 when the prototype Viscount had made it's maiden flight, some thirty-six years before. First and second line airlines were now starting to sell these aircraft to smaller airlines, who were much less likely to contemplate paying us to help them.

Well, business started to come our way. At the Farnborough Show we ran into a couple of chairmen of airlines which took us to jobs in the Midlands and Frankfurt for a start, and over the next five years we were called to Norwich, Copenhagen, Zaire and the Philippines to name some of them.

We kept in close touch with our friends at Rolls-Royce in East Kilbride and it was through them that one day we found ourselves in Kinshasa, Zaire. A Belgian gentleman from that country, who apparently held the position of Aviation Minister, approached Rolls-Royce and asked them if they would send somebody to inspect and assess what was necessary to bring an old YS-11 aircraft back to a state of airworthiness. R-R were not interested in the job as it would involve the airframe as well as the engines, and so they asked us if we would undertake it. We jumped at it then, but today I don't think we would be so keen! We were met at the airport by the Belgian gentleman and two Zairian men in smart suits which displayed the tell-tale armoury bulge at breast pocket level. Throughout the week we were there, these two gentlemen were never very far away from us and looking back, perhaps that was not a bad thing.

The old YS-11 aircraft turned out to have seen better days. To start with, somebody had been very helpful by adding sand into the engine oil tank. The turbine blades were there but only just and to inform our Belgian friend that both engines required a full

*A British Aeropsace 748 Series 2B, D-AHSC, delivered to Deutsche Luft Transport (DLT) in July 1981.*

overhaul did not stretch our technical know-how very far. The airframe was generally in one piece but there were pieces missing here and there and a fair amount of corrosion which was severe in the floor stucture near the toilets. This latter aspect is nearly always present on aircraft which have carried people who are not conversant with using flush toilets. So the airframe would also need an overhaul, all in all a very costly proposition.

During the week while we were clambering around on the aircraft, we were visited each day by a very worried little man who said he was from the Airworthiness Department. Each day he pleaded with us:

*"I hope you are not going to fly, it's not in a suitable condition".*

Each day we tried to put his mind at rest, and added that we were keen to live a little longer!

Well, our inspection only took a couple of days, but we stayed a bit longer to help remove the engines and prepare a written report. This done, we packed our bags and prepared to leave the hotel whose desk staff sought, in fact insisted, that we pay the bill before they waved us a friendly farewell. Unfortunately, our Belgian friend, who had agreed to take care of the bill, had not turned up and while we tried to explain this arrangement the two gentlemen in the bulging lounge suits moved a little closer. Fortunately, after a rather uncomfortable ten minutes, our Belgian friend turned up and all was settled.

We were driven in a Jeep to a Government building and taken to wait in an office, which turned out to be the outer office to President Mobuto's office. Apparently he had to sign or at least approve payment for our week's work and of course he wasn't there. We waited for about half an hour but still no sign of any money for us, and I should imagine we were far from his thoughts - if in fact he had ever heard of us. Our flight home was due to take-off in fifteen minutes and the airport was some five miles away, when we gave up on the money. With our Belgian friend driving the Jeep and Gordon and I hanging on with our eyes tightly closed, we tore into the airport through a back service gate and straight up to the waiting DC10 - no checking in, no passport or security checks, straight up the steps and into our seats. We had hardly sat down when we were taxying out, barely time to even wave goodbye to the bulging lounge suits! Quite an interesting week, we reflected, but financially disappointing although we eventually did get paid by a round-about route. But that is another story!

This period in our lives was most rewarding. We were still in the aircraft business, but in a much smaller way we owned the whole business - the policy decisions, advertising, sales, planning, accounting as well as the technical side, and we were answerable only to ourselves.Unfortunately, after about five years the business did fade away. Our contemporaries, friends and contacts also retired or moved on, and some just passed away.

So as always, all good things come to an end. But there remain so many good memories, some of which we have included in this little book. Hope you have enjoyed it!